The Future of Cultural Tourism

THE FUTURE OF TOURISM

Series Editors: **Ian Yeoman**, *NHL Stenden University of Applied Sciences, the Netherlands* and **Una McMahon-Beattie**, *Ulster University, Northern Ireland, UK*

Some would say that the only certainties are birth and death; everything else that happens in between is uncertain. Uncertainty stems from risk, a lack of understanding or a lack of familiarity. Whether it is political instability, autonomous transport, hypersonic travel or peak oil, the future of tourism is full of uncertainty but it can be explained or imagined through trend analysis, economic forecasting or scenario planning.

The Future of Tourism sets out to address the challenges and unexplained futures of tourism, events and hospitality. By addressing the big questions of change, examining new theories and frameworks or critical issues pertaining to research or industry, the series will stretch your understanding and generate dialogue about the future. By adopting a multidisciplinary perspective, be it through science fiction or computer-generated equilibrium modelling of tourism economies, the series will explain and structure the future – to help researchers, managers and students understand how futures could occur. The series welcomes proposals on emerging trends and critical issues across the tourism industry and research. All proposals must emphasise the future and be embedded in research.

All books in this series are externally peer-reviewed.

Full details of all the books in this series and of all our other publications can be found on http://www.channelviewpublications.com, or by writing to Channel View Publications, St Nicholas House, 31–34 High Street, Bristol, BS1 2AW, UK.

THE FUTURE OF TOURISM: 11

The Future of Cultural Tourism

Edited by
Xavier Matteucci and Simone Moretti

with Licia Calvi and Jessika Weber-Sabil

CHANNEL VIEW PUBLICATIONS
Bristol • Jackson

DOI https://doi.org/10.21832/MATTEU9288
Names: Matteucci, Xavier, editor. | Moretti, Simone, editor.
Title: The Future of Cultural Tourism/Edited by Xavier Matteucci and Simone Moretti.
Description: Bristol; Jackson: Channel View Publications, [2025] | Series: The Future of Tourism: 11 | Includes bibliographical references and index. | Summary: "This book provides multi-layered and nuanced perspectives on how drivers of change may influence cultural tourism on a global, national and local level. As such, it contributes to a greater understanding of how cultural tourism will be governed, performed and experienced within a volatile, uncertain and complex future environment"—Provided by publisher.
Identifiers: LCCN 2024050631 (print) | LCCN 2024050632 (ebook) | ISBN 9781845419288 (hbk) | ISBN 9781845419271 (pbk) | ISBN 9781845419301 (epub) | ISBN 9781845419295 (pdf)
Classification: LCC G156.5.H47 F88 2025 (print) | LCC G156.5.H47 (ebook) | DDC 338.4/791—dc23/eng/20241129
LC record available at https://lccn.loc.gov/2024050631
LC ebook record available at https://lccn.loc.gov/2024050632

Library of Congress Cataloging in Publication Data
A catalog record for this book is available from the Library of Congress.

British Library Cataloguing in Publication Data
A catalogue entry for this book is available from the British Library.

ISBN-13: 978-1-84541-928-8 (hbk)
ISBN-13: 978-1-84541-927-1 (pbk)

Channel View Publications
UK: St Nicholas House, 31–34 High Street, Bristol, BS1 2AW, UK.
USA: Ingram, Jackson, TN, USA.

Authorised Representative: Easy Access System Europe – Mustamäe tee 50, 10621 Tallinn, Estonia gpsr.requests@easproject.com.

Website: https://www.channelviewpublications.com
Bluesky: https://bsky.app/profile/channel-view.bsky.social
X: Channel_View
Facebook: https://www.facebook.com/channelviewpublications
Blog: https://www.channelviewpublications.wordpress.com

Copyright © 2025 Xavier Matteucci, Simone Moretti and the authors of individual chapters.

All rights reserved. No part of this work may be reproduced in any form or by any means without permission in writing from the publisher.

The policy of Multilingual Matters/Channel View Publications is to use papers that are natural, renewable and recyclable products, made from wood grown in sustainable forests. In the manufacturing process of our books, and to further support our policy, preference is given to printers that have FSC and PEFC Chain of Custody certification. The FSC and/or PEFC logos will appear on those books where full certification has been granted to the printer concerned.

Typeset by SAN Publishing Services.

Contents

	Figures and Tables	vii
	Contributors	xi
	Acknowledgements	xvii
1	The Future of Cultural Tourism: Utopia, Dystopia, Heterotopia *Xavier Matteucci*	1

Part 1: Governance

2	Communities of Hospitality as a Democratic Governance Model for Host–Guest Interactions *Prosper Wanner, Hakan Shearer Demir and Ivana Volić*	19
3	Envisioning Posthumanist Cultural Tourism: Indigenous Sociomaterial Practices of Teaching Tourists about Local Cultures *Ella Björn and Monika Lüthje*	34
4	Fostering Sustainable and Resilient Rural Communities through Cultural Tourism Villages: A Case Study of the Dalmatian Hinterland *Lidija Petrić, Ante Mandić and Davorka Mikulić*	50
5	Uncovering the Multifaceted Heritage Values of Longhushan World Natural Heritage Site through Tourists' Lens: An Analysis of Online Travelogues *Rouran Zhang, Weili Zhan, Ying Lyu and Da Kuang*	70

Part 2: Consumption

6	Edifying Slow Cultural Tourism Concepts and Practices *Jelena Farkić*	91
7	Emerging Perspectives on the Future of Cultural Tourism *Dallen J. Timothy*	107
8	Museums of the Future: Cultural Tourism Experiences for Wellbeing and Transformation *Marta Šveb Dragija and Daniela Angelina Jelinčić*	121

Part 3: Technologies

9 Cultural Tourists of the Future: Envisioning How Digital Technologies Can Shape Heritage Experiences in Europe 139
Emanuele Mele

10 Beyond Digital Prisons: Counterculture and the Hippie Trail Resurrected 155
Daniel W.M. Wright

11 Cultural Tourism in the Metaverse 170
Ulrike Gretzel and Eva Sánchez-Amboage

Part 4: General Outlook

12 Quality of Life and Cultural and Heritage Tourism 189
Muzaffer Uysal and Jiahui Wang

13 Cultural Tourism as Cultural Adaptation: Urban Design Scenarios for the Future 207
Maurizio Scarciglia

14 Uncharted Territories in Cultural Tourism: Synthesis and Some Personal Reflections 227
Simone Moretti and Xavier Matteucci

Index 243

Figures and Tables

Figures

Figure 3.1	Practice categories of teaching about Sámi cultures in tourism	44
Figure 4.1	Rural villages' clusters of population	55
Figure 4.2	Mapping of the population-based clusters of villages in the Split-Dalmatia hinterland	56
Figure 4.3	Veliki Godinj – an 'open-air museum'	60
Figure 4.4	Grabovci (Proložac Gornji) – 'diffused eco-hotel'	61
Figure 4.5	Ethno village Kokorići – CBT village for culturally immersed tourists	61
Figure 4.6	Zagvozd – a lighter version of CBT village for culturally immersed tourists	62
Figure 4.7	Stella Croatica – non-authentic profit-oriented model of a tourist village	63
Figure 5.1	World Heritage Nomination Process	73
Figure 5.2	Semantic network analysis diagram	79
Figure 9.1	World Heritage Journeys of Europe [Homepage online image]	147
Figure 9.2	Three cultural tourist types of a utopian future	150
Figure 11.1	The role of technology in shaping cultural tourism practice	172
Figure 11.2	Technology use in museums	174
Figure 12.1	Temporal network	191
Figure 13.1	Villa Adriana in Tivoli	211
Figure 13.2	Concept image of the Palm Jumeirah global reproduction and the spread of the Dubai resort city model	214

Figure 13.3	Concept image of urban renovation interventions in the historical centre of Rovereto	217
Figure 13.4	Concept image of the interface of 'Stay in the Doughnut': An urban game to stimulate citizens' participation in curating the cultural program of Rovereto	218
Figure 13.5	Concept image of Piazza Rosmini: The countryside in the city (automated agricultural process)	219
Figure 13.6	Concept image of Piazza Rosmini: Creation of a catering hub for the Trentino cuisine festival	220
Figure 13.7	Concept image of Piazza delle Oche: Mushrooms farming with recycled coffee powder	221
Figure 13.8	Concept image of MART Museum: Hologram reconstruction of artwork from international collections	221
Figure 13.9	Concept image of MART Museum: The museum returned to nature – landscaping installation by Olafur Eliasson	222
Figure 13.10	Concept image of Piazza delle Erbe: Open Air Cinema (RAM Festival of Cinema and Archaeology)	223
Figure 13.11	Concept image of Piazza delle Erbe: International Congress on Robotic Surgery	223
Figure 13.12	Concept image of the Leno stream: Rovereto Music Festival on the stream banks	224
Figure 13.13	Concept image of the Leno stream: A resilient landscape integrating historical apple cultivation from Trentino Alto Adige	224
Figure 14.1	Model of possible cultural tourism futures	229

Tables

Table 4.1	List of protected rural ensembles in the Dalmatian hinterland	57
Table 4.2	Models/scenarios of tourism villages in the Dalmatian hinterland	58
Table 5.1	Analysis on the authoritative Evaluation of the Longhushan Heritage	74

Table 5.2	Analysis of the value of Longhushan heritage evaluated by Chinese scholars	75
Table 5.3	Review data obtained by travel websites	76
Table 5.4	Semantic classification of high-frequency words	77
Table 5.5	Heritage value classification based on high-frequency words	77
Table 10.1	Digital ages	160
Table 11.1	Typology of immersive technology experiences	179
Table 12.1	QoL and cultural and heritage tourism research: Tourist and resident perspectives	196
Table 12.2	Current research topics and future areas in the studies of cultural tourism and QoL	199
Table 14.1	Key concepts and implications for cultural tourism futures	237

Contributors

Ella Björn holds a master's degree in tourism research and works as a project planner in the Faculty of Art and Design at the University of Lapland, Finland. Her research focuses on service design, art-based methods and placemaking in tourism.

Licia Calvi was educated as a Digital Humanist. She is a senior member of the Digital Transformation of Cultural Tourism research group at the Academy for Tourism at the Breda University of Applied Sciences and a member of the Executive Committee of the Experience Research Society. She has been involved in various consultancy projects related to the use of storytelling for heritage. Next to her interest in using storytelling as a design tool for experiences, she is also interested in understanding the use and impact of technology, in particular of extended realities, for the design of sustainable and authentic experiences around cultural heritage.

Jelena Farkić holds a Lecturer/Researcher position at Breda University of Applied Sciences. She is an active Forest Therapy South Eastern Europe member and is its certified forest wellness mediator. Jelena has been exploring wellbeing through concepts and practices such as slow adventure, forest bathing (shinrin-yoku) or idleness, underpinning her more recent work with relational philosophies. Her studies offer critical insights into the emergent interactions between humans and non-humans within diverse tourism spaces, while she keeps wondering how such dynamics might sustain multi-species communities and contribute to planetary wellbeing.

Ulrike Gretzel is a Senior Research Fellow at the Center for Public Relations, University of Southern California and Lecturer at the IMC University of Applied Sciences Krems. Her research spans the design, use, and implications of emerging technologies, ranging from social media and mobile applications to smart cities, robots, and the Metaverse. She has over 20 years of experience conducting academic and practice-focused research. She is frequently acknowledged as one of the most cited authors in the fields of tourism and persuasion and is an elected fellow of the International Academy for the Study of Tourism.

Daniela Angelina Jelinčić is a Senior Research Advisor/Full Professor employed by the Institute for Development and International Relations (IRMO), Croatia. She holds a PhD in Ethnology from the University of Zagreb. Her specific interests are in cultural tourism, cultural heritage management, cultural/creative industries, cultural policy, creativity, experience economy and social innovations. As an adjunct professor, she teaches at various institutions in Croatia, and occasionally at the Helsinki University of Arts and at the UNESCO Chair for Cultural Heritage Management and Sustainable Development, Institute for Advanced Studies (iASK) in Köszeg, Hungary. She has authored and edited a number of publications and she has served as an expert for cultural tourism at the Council of Europe. She has also coordinated and participated in a number of interdisciplinary research projects.

Da Kuang is an Assistant Professor in the School of Architecture and Urban Planning, Shenzhen University, China.

Monika Lüthje works as a Senior Lecturer in Tourism research at the University of Lapland, Finland. Her research focuses on cultural sensitivity and Indigenous tourism.

Ying Lyu is a postgraduate candidate in the School of Architecture and Urban Planning, Shenzhen University, China.

Ante Mandić is an Assistant Professor of Sustainable Tourism at the University of Split in Croatia and he is also affiliated with Colorado State University. He is Associate Editor of the Journal of Ecotourism and serves as a coordinator of knowledge development at the International Union for Conservation of Nature's TAPAS. His research focuses on sustainable tourism in nature-based destinations, and he has authored books with Springer Nature, Routledge, and Edward Elgar. He also has extensive experience in international projects with INTERREG MEDITERRANEAN and HORIZON 2020.

Xavier Matteucci is an independent scholar teaching at different universities in Austria and beyond. He holds an Honorary Professorship at IMC Krems University of Applied Sciences. His publications cover the areas of tourism experiences and well-being, cultural tourism and qualitative social research methodologies. He is particularly interested in tourism as a space for exchange, learning, activism, and as a way to reclaim control over our lives. His most recent work includes *The Creative Tourist: A Eudaimonic Perspective* (Emerald Publishing, 2024). Apart from his academic activities, Xavier has led European Commission funded consultancy research projects on sustainable cultural tourism in South-East Europe.

Emanuele Mele is Scientific Collaborator at the Institute of Tourism at the University of Applied Sciences and Arts Western Switzerland (HES-

SO), in Sierre. He is also External Research Associate at USI – Università della Svizzera italiana, the UNESCO chair in ICT to develop and promote sustainable tourism in World Heritage Sites, *Lugano, Switzerland*. He holds a PhD in Communication Sciences. His research interests focus on digital marketing in tourism and hospitality, cross-cultural communication, and the use of information and communication technologies for heritage promotion.

Davorka Mikulić is an Assistant Professor at the Faculty of Economics, Business and Tourism, University of Split. Her main research interests lie in culture heritage management, cultural tourism, destination marketing and the economics of travel and tourism. She has been teaching in higher education for more than 15 years. She has published scientific papers in the areas of cultural tourism development and destination marketing. She has also been involved in national and international research projects.

Simone Moretti is a Researcher and Lecturer in Tourism at Breda University of Applied Sciences, in the Netherlands. After his studies in Business and Economics and some work experiences in the field of business development, Simone completed a Master in Tourism Destination Management. He continued his career in the tourism industry, working as a project manager in Thailand and Belgium. His research now focuses on sustainable tourism development, cultural tourism and tourism impacts on society. His current research activities include several EU-funded projects, such as SmartCulTour (Smart Cultural Tourism as a Driver of Sustainable Development – Horizon 2020) and Tourban (Boosting sustainable tourism development and capacity of tourism SMEs – COSME).

Lidija Petrić is a Full Professor at the Faculty of Economics, Business and Tourism, University of Split, Croatia. Her primary research areas are tourism economics, tourism development, destination management and planning and cultural tourism. She has contributed to numerous international, EC-funded and local projects and worked with esteemed organisations like UNDP, GEF, WWF, and UNEP. Professor Petrić is a member of the European Regional Scientists Association, the Scientific Committee for Tourism at the Croatian Academy of Sciences and Arts, and the Croatian UNESCO National Committee.

Eva Sánchez-Amboage is an Assistant Professor in the area of Marketing and Market Research at the University of A Coruña, where she teaches in the Business Administration and Management, Audiovisual Communication degrees, as well as in the Master's program in Planning and Management of Tourist Destinations and Products. She serves as an editor for Redmarka: Academic Journal of Applied Marketing and is a researcher in the iMARKA research group at the University of A Coruña. Her research primarily focuses on tourist communication in social media, new technologies applied to tourist marketing, and influencer marketing.

Maurizio Scarciglia is an Architect and Urbanist. He is a Lecturer and Researcher of Urban Design at Breda University of Applied Sciences in The Netherlands. Since 2007, Maurizio is the founding director of NAUTA architecure & research, an office based in Rotterdam, operating in the fields of Urbanism and Cultural Studies.

Hakan Shearer Demir specialises in human rights and democratisation, with a specific focus on sustainable development, displacement, local governance and cultural heritage. Dr Shearer Demir has worked with non-governmental organisations as well as inter-governmental organisations, including the Council of Europe and several United Nations agencies. He has an MA in International and Intercultural Management and a PhD in Political Science. He is currently an adjunct faculty member at the University of Strasbourg, and his recent work has focused on heritage inclusion, community regeneration and governance, with particular attention to the co-construction of communities after displacement.

Marta Šveb Dragija is a Research Assistant employed by the Institute for Development and International Relations (IRMO), Croatia. She holds a MSc degree in psychology from the University of Groningen (NL) where she is also currently finishing her PhD in psychology. As an Assistant Professor, she teaches at the Edward Bernays University of Applied Sciences in Zagreb. Marta is a member of the European Association for Social Psychology (EASP). She has authored several scientific articles and startegic documents. Her research interests are in cultural tourism, experience economy and wellbeing.

Dallen J. Timothy is Professor of Community Resources and Development, and Senior Sustainability Scientist at Arizona State University. He is also a visiting professor at universities in China, Spain and Mexico and Research Associate at the University of Johannesburg, South Africa. He has ongoing research projects in various parts of the world in the areas of international borders and the geopolitics of tourism, heritage management, pilgrimage and tourism issues in the Global South.

Muzaffer Uysal is a Provost Professor of the Department of Hospitality and Tourism Management – Isenberg School of Management at the University of Massachusetts, Amherst, USA. His current research interests centre on tourism demand/supply interaction, impact and tourism development and quality-of-life research in tourism and hospitality.

Ivana Volić is a tourism and culture researcher, lecturer and consultant. She holds a master's degree in Tourism Studies, and a PhD in Cultural Management and Policy. Her research interests include cultural and tourism policy, cultural heritage, community development and participative planning. She is currently an adjunct faculty at IMC Krems University of

Applied Sciences, in Austria. Ivana is also a member of the Faro Network of the Council of Europe. She has participated in various EC funded projects on issues of cultural heritage and tourism.

Jiahui Wang is a PhD student in Hospitality and Tourism Management – Isenberg School of Management at the University of Massachusetts, Amherst, USA. Her research interests include quality of life, tourist behaviour and dark tourism.

Prosper Wanner is a project management engineer and sociologist. He is Associate Professor at the University of Aix-Marseille in cultural mediation of the arts. Wanner is one of the lead managers of the cooperative of inhabitants *Hôtel du Nord* in Marseille and a researcher at the cooperative platform *Les oiseaux de passage*. He is also an expert at the Council of Europe on the value of cultural heritage for society. Wanner's doctoral thesis examined the nexus between hospitality, tourism, digital platform and human rights at the University of Paris-Nanterres.

Jessika Weber-Sabil is Professor of Digital Transformation in Cultural Tourism at the Academy for Tourism of Breda University of Applied Sciences (NL). Jessika holds a PhD from Bournemouth University (UK), where she explored the mobile game experience of tourists with location-based augmented reality games in urban environments. Among other projects, from 2016 to 2022, she worked as a senior researcher and project manager under the Professorship of Serious Games, Innovation & Society where she led research projects on games and tourism.

Daniel W.M. Wright has a PhD in post-disaster tourism management and development. He also publishes widely in the subject area of tourism futures. His research and journal publications explore some of the more challenging issues facing the future of tourism. His research covers wider societal, environmental and technological matters. His publications have also attracted wider international media attention.

Weili Zhan is a postgraduate candidate in the School of Architecture and Urban Planning, Shenzhen University, China.

Rouran Zhang is a Research Associate in the McDonald Institute for Archaeological Research at the University of Cambridge, UK, and an Associate Professor and Vice Director of the Department of Landscape Architecture in School of Architecture and Urban Planning at Shenzhen University, China. He is the Vice President of ICOMOS International Cultural Tourism Scientific Committee. He holds a PhD in Interdisciplinary Cross-Cultural Research from the Centre of Heritage and Museum Studies, Australian National University. He is the author of *Chinese Heritage Sites and Their Audiences: The Power of the Past* (2020), published by Routledge.

Acknowledgements

We would like to express our gratitude to the authors who have generously contributed to this edited volume. This book would have never come to light without their dedication, patience and scholarship. Some of the contributing authors have thoroughly reviewed some chapters. Thank you for that. Our thanks also go to the book series editors Ian Yeoman and Una McMahon-Beattie, to Sarah Williams and the editorial and production teams at Channel View who have been very patient, supportive and responsive at all times.

Xavier Matteucci is grateful to his close family and friends for their unconditional love and support. Thank you also to Chris Rout and Bernat Corominas-Murtra for our quality intellectual and inspirational conversations. Special thanks go to Miriam and Olivier at Banc Public for affording me the opportunity to enter a heterotopic oasis in times of sociopolitical antagonism. Lastly, thank you Simone for bearing with me throughout the editing process.

Simone Moretti is immensely thankful to his family for their constant support, especially to Bharty for her love and understanding. Thank you also to Xavier for his commitment and our intriguing discussions on the future of cultural tourism. Special thanks to all the places I have been able to visit in the last few years, their people, and communities. You have truly been an endless source of reflection and inspiration on the current and future meaning of culture, heritage and tourism.

1 The Future of Cultural Tourism: Utopia, Dystopia, Heterotopia

Xavier Matteucci

The future of cultural tourism is intricately dependent on the future of humanity. This statement is a truism and is indisputable. Like tourism, which is a social phenomenon, cultural heritage is the product of humans' interaction with the sociomaterial environment. The concept of *cultural tourism* has no clear boundaries. However, cultural tourism is usually understood as a commercial phenomenon in which tourists with specific interests experience a destination's cultural heritage assets. These assets can be vast and diverse such as monuments, archaeological sites, museums, architectural complexes, festivals, rituals, customs, sports (traditional games), traditional crafts, languages, artistic practices (e.g. music, dance, literature, films), cultural landscapes and the everyday lives of the local population. Envisioning the ways tourists and residents of tourism destinations will experience, consume, understand, produce and articulate cultural heritage in the future is, therefore, dependent on the state of humanity, which may or may not change in the short or long term. While essential as a topic, the future of humanity is too complex and manifold, and such an analysis is beyond the scope of this edited volume. However, the future of humanity is a topic that cannot be ignored. It cannot be ignored because thinking and anticipating the future can guide individuals and organisations, such as governments, to manage the present (Urry, 2016). In a seminal piece published in 2009, Swedish philosopher Nick Bostrom asks fundamental questions such as

> whether and when Earth-originating life will go extinct, whether it will colonize the galaxy, whether human biology will be fundamentally transformed to make us posthuman, whether machine intelligence will surpass biological intelligence, whether population size will explode, and whether quality of life will radically improve or deteriorate. (2009: 186)

Given the context of fast-paced technological advancement in which we live, these are all the more legitimate and important questions. If technological innovation has been the main driver of economic development (Bostrom, 2009), it has also contributed to climate change, massive extraction and depletion of natural resources, destruction of wildlife habitat, extinction of Earth's species, international insecurity, nuclear threat and increased population surveillance. It is, therefore, unquestionable that the future consumption and production of cultural tourism will depend on our capacity to learn from experience and our ability to anticipate possible outcomes in the event that current trends remain stable. Based on Bostrom's fundamental questions, for instance, we may speculate that if humans colonise the galaxy, new forms of cultural tourism will emerge. Space tourism had long remained in the realm of science fiction or fantasy; however, since the first paying space tourist travelling on a Russian Soyuz rocket in 2001, it has now become a reality (Toivonen, 2022) and further space travel development is underway with the Moon and Mars as destinations (Spennemann, 2007). For instance, if humans develop into posthumans, we may expect enormous population growth, greater life expectancy and enhanced cognitive and physical capacities, which may provide opportunities for new forms of cultural consumption and production, but may also add pressure on cultural industry workers and on some cultural heritage assets.

With the implementation of intelligent automation, for instance, millions of both low-skilled and specialised jobs are predicted to disappear (Tussyadiah, 2020). Robots will work together with humans serving tourists (Ivanov, 2023), and robots' performances may become indistinguishable from the performances of humans (Yeoman & McMahon-Beattie, 2020). If machine intelligence surpasses biological intelligence, we may see a future where a few powerful humans and/or despotic governments will control all other humans (Bostrom, 2019) or perhaps we may even see a future where robots are customers (Ivanov, 2023) and where machines overrun humans. In such dystopian scenarios, cultural tourism would be radically altered.

While a degree of uncertainty about the future will always prevail, there seems to be a consensus among scientists that the two most likely and dramatic types of risk to human survival, besides a super volcanic eruption, are nuclear disasters and environmental threats (e.g. global warming), both caused by human activities. In the event of societal collapse, cultural tourism, as a social phenomenon, would suddenly appear irrelevant to most of us. Indeed, as Wright (2023: 152) argues, 'if a doomsday scenario became reality, humans across the globe would be impacted differently', yet potentially forcing many of us 'to live and travel like that of our hunter-gather ancestors'. While a major existential disaster is plausible, it is unclear when this may happen and to which extent it may impact us around the world. In a recent article (Matteucci *et al.*, 2022a), we have theorised three likely futures for cultural tourism, which we describe as utopian, dystopian and heterotopian. These three futures stem from an optimistic and Western-centric stance in which we envision a privileged future in which (cultural) tourism still somehow has a place. It goes without saying that a utopian future is preferable; however, as Urry (2016: 11) warns, 'What is preferable may turn out to be the least probable'.

Utopian Future

Utopia is commonly understood as an ideal world, or a sociopolitical system in which all forms of life live in peace and harmony. Within this utopia, diverse voices, values and cultures are expressed and acknowledged in order to arrive at a well-informed and democratic consensus. In a utopian future, cultural tourism governance would pursue municipalist, degrowth strategies underpinned by post-anthropocentric values. Post-anthropocentrism entails that both human and non-human forms of life are equally, intrinsically valuable and interdependent (Benson, 2019). Furthermore, degrowth does not only mean reducing the throughput of human economic activities (such as tourism), but it also offers alternative (non-economic) visions of development (Sharpley & Telfer, 2023). Degrowth and a circular economy would allow us to envision a stable future 'via cyclic regenerative processes' (Tomassini & Cavagnaro, 2022: 344). Manifestations of regenerative processes could be rejuvenating local heritage and rehabilitating abandoned real estate for the benefit of locals and visitors (Tomassini & Cavagnaro, 2022). Municipalism similarly contests austerity and capitalist exploitation of humans and non-humans. Municipalism represents an alternative political thrust to 'democratise society' through a politics of proximity, solidarity-making and self-organising autonomous grass-root networks (Thompson, 2021). A utopian cultural tourism future would witness greater protection of heritage assets, including wildlife and the natural environment. It would also mean that decisions about heritage management (e.g. preservation and interpretation) would follow the governance principles of participatory democracy, which is an intrinsic facet of (new) municipalism, and which, according to Schaap and Edwards (2007), entails that:

> people have equal right to liberty and self-development, which can only be achieved in a society that fosters a sense of political efficacy, nurtures a concern for collective problems, and contributes to the formation of a knowledgeable citizenry capable of taking a sustained interest in the governmental process. (2007: 664)

From the perspective of participatory democracy, not only dialogical interactions should involve a diverse set of community stakeholders, including marginalised groups, but also the interests and rights of local residents, as heritage bearers, would be prioritised over those of tourists and of foreign investors (Higgins-Desbiolles *et al.*, 2019; Matteucci *et al.*, 2022b). This is not to dismiss the role of tourists in heritage-making; rather, as Smith (2015: 139) argues, heritage, whether intangible or material, can only be protected if 'it is used, and made meaningful, in the context of contemporary needs and aspirations of the communities to whom it is significant'. Morin (1999: 58) reminds us that:

> democracy expects and nurtures diversity of interests and diversity of ideas. Respect for diversity means that democracy cannot be confused with dictatorship of the majority over minorities; it must include the rights of minorities and protesters to exist and express themselves; it must allow the expression of heretical and deviant ideas. Just as the diversity of species must be protected to safeguard the biosphere, so the diversity of ideas, opinions, information sources and media must be protected to safeguard democratic life.

From a utopian perspective, cultural heritage is understood as a democratic process or performance that is enacted and managed through subjective political negotiation of identity, place and memory by community groups who stand as heritage bearers (Smith, 2015). This view of cultural heritage resonates with trends in cultural tourism scholarship that emphasise the fluid processes of heritage-making and place-making, which are illustrated in Matteucci and von Zumbusch's (2020) recent definition of cultural tourism as:

> a form of tourism in which visitors engage with heritage, local cultural and creative activities and the everyday cultural practices of host communities for the purpose of gaining mutual experiences of an educational, aesthetic, creative, emotional and/or entertaining nature. (2020: 19)

Place-making reflects the idea of people's appropriation of places (including local heritage therein), yet, in a non-exploitative way. Richards (2020) argues that place-making can facilitate regenerative processes through creativity and transformative practices and encounters. Place-making can be enacted through various forms of slow tourism mobilities, and arguably through digital travel (or virtual tourism). Slow mobilities include low-carbon footprint leisure, proximity tourism and staycation. The notion of slowness is highly subjective (Matteucci & Tiller, 2023); it connotes ideas of sustainability (Tzanelli, 2021) and involves seeking quality over quantity, connecting with people, places and cultures and supporting local small-scale producers, farmers and retailers (Slow Movement, 2024). The germane concept of Slow City (Cittaslow), which gives prominence to local history, is used to promote local distinctiveness and sustainable urban development (Robinson *et al.*, 2020). In the same vein, proximity tourism (or travel near home) is said to generate many benefits to communities such as invigorating endogenous cultural heritage, and reinforcing local cultural identities (Arrieta Urtizberea *et al.*, 2016). Proximity tourism has also been referred to as *locavism* by Houge Mackenzie and Goodnow (2021) who assert that micro-adventures pursued close to home bring valuable social and economic benefits to peripheral rural areas. Slow (walking) adventures have also been associated with well-being outcomes to those who partake in them (e.g. Farkić & Taylor, 2019).

While digital travel presents some clear advantages in terms of its rather low-carbon footprint, and in terms of inclusivity (ease of access to cultural experiences for those with impaired mobility) (Tzenalli, 2021), the psychological and sociocultural benefits of digital travel to communities are less obvious. Furthermore, a utopian cultural tourism future is one in which technological advances in genetics, nanotechnology and robotics are strictly controlled to serve humanity to preserve cultural heritage and the nature of human life. In short, slow mobilities, proximity tourism, creative tourism, and to a lesser extent virtual tourism, are often depicted as soft alternatives to mass cultural tourism.

From the optimistic lens of *hopeful tourism* (Pritchard *et al.*, 2011), which entails co-transformative learning and action, 'cultural creatives' are presented as the vital agents who will be driving positive sociocultural change (Ateljević, 2020). How many well-intentioned cultural creatives will it take to mobilise change through transformative action? While some inclusive and regenerative cultural tourism initiatives are to be found around the world, capitalism – as a form of civilisation – is resilient (Aubenas & Benasayag, 2002; Tzanelli, 2021), and it seems that a sheer lack of political will is holding democratic, regenerative endeavours back. This lack of political commitment to making Earth a cleaner, safer, healthier and more ethical place to live compels us to anticipate many disruptions.

Dystopian Future

The opposite of utopia is dystopia, which refers to a frightening society or 'a futuristic anti-utopia' (Daniels & Bowen, 2003: 423). As we observe world politics around us, it is hard not to despair. In fact, a thorough analysis of US death statistics reveals that despair, as a form of anxiety disorder, is associated with a growing number of deaths in the United States (Case & Deaton, 2020). Professors of economics at Princeton University, Anne Case and Angus Deaton note that, in the last decades, middle-aged, blue-collar workers have increasingly been suffering from alcoholism, substance overdose and suicide (depression), all of which they link with the decline in wages and in the quality of jobs, with the deterioration in family relationships and with feelings of social exclusion. This societal malaise is similarly observed by leisure researchers Kumm and Pate (2024: 396):

> As we survey our contemporary moment, as well as our own feelings and emotions, we are dismayed at the relentless diminishing of people's

powers of existence. In the United States, we witness the sad affects in the banning of books, speech, and ideas – especially those that propagate the truth of racial, sexual, gender, and class injustice. We also see the removal of rights to bodily autonomy, privacy, and health care. Environmental protections are eroding; wages are suppressed; markets are manipulated; workers are exploited. We see widespread and open dissemination of lies and propaganda leveraged for political control. Mirroring global conflicts and struggles, sadness is all around.

Despair is no longer the curse of the poor, now despair is also spreading among Western, richer societies. Despair is everywhere. As recently evidenced in Case and Deaton's work, the destructive force of neoliberal policies (or savage market capitalism) is held responsible for growing inequalities, precarity, poverty and for the predicament of global warming (Braidotti, 2019; Chomsky & Pollin, 2020; Monbiot, 2016). According to Herrington (2023: 3), 'countries that continue to chase growth while the ecosystem breaks down are heading for disruption'. Despair, caused by job loss, social isolation, and loss in meaning in life, is also an unfortunate consequence of the COVID-19 crisis (Pies, 2020). In that respect, Seedhouse (2020) contends that during the recent pandemic, governments refrained from inviting their populations to democratically debate policies like lockdown; instead, politicians failed to understand the meaning of scientific evidence and resorted to propaganda to terrify citizens into compliance. The current conflict in Ukraine and the appalling mass killing of civilians in the Gaza Strip (ICJ, 2024) can only accentuate feelings of anger, hopelessness and despair. These recent violations of basic human rights, international law and the rise of conservative politics in Western 'democracies' intimate a rather bleak future.

Within this distressing context, we articulated a dystopian vision of cultural tourism in which local communities and their endogenous heritage have become increasingly commodified to meet the corporate thirst for growth and quick profit. Sites of cultural significance and public spaces have been privatised and heavily commercialised (see Zorzin, 2015 on the privatisation of archaeological sites). For instance, Russo and Scarnato (2018) have well documented the touristification process of Barcelona that they associate with the 'growth machine' discourse of neoliberal politics, and which resulted in social segregation, higher cost of living, the expropriation of public space, the privatisation of heritage assets and the homogenisation of the cityscapes with invasive global brands. Mass cultural tourism and the touristification of local economies, culture and society have made it increasingly difficult to appreciate what is local and traditional (Richards, 2023). Not only have most locals and migrant workers been pushed to the periphery of town centres where real estate prices are lower, but they have also become subservient to tourism operators and tourists. The gap between the rich and the poor has continued to grow to such a point that more and more potential tourists are unable to afford to take any kind of holiday at all; and in the event that they can, tourists respond to prices rather than to quality cultural offers.

A dystopian future of cultural tourism envisions an alienated, mostly proletarian, and consumerist society, in which cultures have become spectacles devoid of deeply rooted meanings, and in which cultural heritage embodies a vision of the world that has been objectified by a powerful and hegemonic elite (Debord, 1967). In *La Societé du Spectacle*, written in 1967, Debord offers a Marxist critique of modern societies that has disturbingly remained pertinent until today. Here, Debord (1992) understands *spectacle* as both the propaganda device of capitalism, and 'social relationship between people mediated by images' (1992: 16). Heritage as spectacle is exemplified in du Cros and McKercher's (2015) marketing recipe for creating cultural tourism attractions, which includes mythologising a cultural asset, emphasising its otherness, making it triumphant, a spectacle, a fantasy, and making heritage fun, light and entertaining. In the Austrian capital, Time Travel Vienna is a visitor attraction that epitomises the 'spectaclisation' of heritage, as suggested by du Cros and McKercher. Time Travel Vienna (www.timetravel-vienna.at/en/) is presented as a unique attraction that invites people to explore the moving history of Vienna 'with 5D cinema, VR-glasses, animatronic wax figures, rides and multimedia shows'. While there is nothing wrong with fostering learning through entertaining tactics, it becomes ethically challenging when Vienna's darkest history is presented as trivial and fun. Here, the embarrassment that Viennese citizens embraced the Nazi regime is astoundingly circumvented with state-of-the-art technical gadgets. Turning 'guilty landscapes' (Reijnders, 2009: 175) into 'anodyne commodities' (Tzanelli, 2021: 383) is an artifice of the past, the present and the dystopian future.

In a dystopian future, heritage, as spectacle, remains an instrument of power that is used to impose historical and cultural narratives of nationhood, citizenship and nationalism (Hollinshead, 2009; Smith, 2015). Smith (2015) refers to authorised heritage discourse (AHD) to describe heritage that is presented as tangible, non-renewable and vulnerable by a small but dominant group of policymakers and so-called professional heritage experts. Smith (2015) contends that:

> [w]ithin the AHD, these experts are defined as the custodians of the human past, whose professional duty it is to not only safeguard but to also provide stewardship for the way the value of heritage is communicated to and understood by non-expert communities. (2015: 135)

While the AHD is only one discourse among others, it is presented (or dictated) as legitimate, thus impinging upon communities' interests, alternative narratives and different cultural identities. In fact, misleading representations of heritage and host communities can lead to tourists' misunderstanding and antagonism (Suleman & Qayum, 2017). Even worse than the threat of misrepresentation, heritage assets, such as sacred sites, may be destroyed during military conflicts (Raj & Griffin, 2017). A dystopian portrayal of the likely future of cultural tourism seems to obliterate the many struggles that are taking place worldwide. However, this vision of the future may be overly pessimistic. Promises of a more ethical and egalitarian future can be found in emancipatory struggles, such as those of communities seeking to be heard and gain greater control over how their local heritage is managed and represented.

Heterotopian Future

In Matteucci *et al.* (2022a), rather than simply utopian or dystopian, we argued that the future of cultural tourism would more likely be heterotopian. Informed by Foucault's (1986) theorisation of heterotopias as counter-sites or sites of resistance, we suggested that a heterotopian perspective would better account for the struggles of many groups who strive to be recognised as heritage bearers within their own localities or regions. Because heterotopias, whether real or virtual, relate to spaces 'in which a utopian vision of the world can be enacted' (Matteucci *et al.*, 2022a: 6), we argued that the future of cultural tourism would manifest itself as pockets of resistance – or bubbles of ethical consumption and practices – within a slowly decaying neoliberal political order.

Such spaces already exist in the form of heritage communities, cooperatives (see Chapter 2) and social enterprises, all underpinned by an ethics of care and the pursuit of the 'common good'. Heterotopian cultural tourism would also include spaces of slow and creative practices such as those offered by local artists and artisans. Based on endogenous resources, handicraft production, such as soap making, leather crafting, glass blowing, rug making, calligraphy, perfume making and olive oil making, not only relies on local knowledge and practices but also caters to the needs of local communities and fosters shared social capital (Scherf, 2021). Through such cultural practices, visitors and locals meet, exchange, collaborate and establish new networks and inspire each other. For their emancipatory power, Matteucci and Smith (2024) have likened creative, cultural tourism spaces to pockets of resistance:

> The creative tourist space is a site of actions, practices and performances; it is a space of experimentation; it stands for a bold rejection of inertia. This way, the creative tourist experience is a manifestation of empowerment, of making one's own freedom. (2024: 93)

The term 'pockets of resistance' not only intimates marginality and nonconformism, but also rebellion, activism, transgression, dissidence and acts of civil disobedience (cf. Thoreau, 1849). In fact, more than 35 years ago, Krippendorf (1987: 107) had already called for 'rebellious tourists and rebellious locals' to resist the iron hand of conservatives who were determined to maintain business-as-usual practices. While Krippendorf may have been a visionary tourism scholar, long before him, many were the dissident voices in other disciplines. Henry David Thoreau (2017: 8 [1849]), for instance, urged women and men not 'to resign [their] conscience to the legislator' when the legislator fails to enforce justice. In his essay *Le Droit à la Paresse [The Right to Be Lazy]*, Paul Lafargue (2023 [1880]) demystifies the dogmatic depiction of work (overproduction) as a virtue to portray it as the cause of all intellectual degeneration and organic destruction. Lafargue advocated three hours of work a day; the rest of the day, he argued, could be devoted to leisure activities. Still today, Lafargue's position may connote utopia to some (e.g. blue-collar and service industry workers) and dystopia to others (e.g. corporation shareholders and plutocrats). Whether such ideas are coherent or nonsensical, past and present dissident voices appear fruitful inasmuch as they can help us imagine alternative futures. Thinking about the future of cultural tourism as heterotopia presses us to question the ways of our current world. It is a form of social critique.

In 1978, American pioneer of information technology and philosopher Ted Nelson noted that political issues had become increasingly technical. He contended that 'the guys who send rockets into space are not considered scientists by the *real* scientists. They're technologists' (1978: 53). Following this line of reasoning, I would argue that, during the COVID-19 crisis, the guys who forced millions of children (and adults) into lockdowns, sending them into social isolation and sadness, and the same or other politicians who made vaccination indiscriminately mandatory to everyone above 18, and who imposed lockdowns on unvaccinated people (e.g. in Austria), were not only little concerned with violating fundamental human rights (ethics), but were also not guided by *real* science (Perronne, 2021; Seedhouse, 2020). Many scientists in positions of power are rarely free of conflicts of interest (see Sismondo, 2021 on corruption in the medical sciences), and politicians... Well, as Kary Mullis (1998), Nobel Prize winner in Chemistry for his invention of the PCR test, asserted:

> politicians don't know anything about scientific things. They just want to look like they do. Somebody has to advise them. Who are those advisors? It is an important question because those people – who are always having to come up with the imminent disasters that can be prevented by governmental projects, sponsored by informed and well-meaning politicians – are manipulating you. They are parasites with degrees in economics or sociology who couldn't get a good job in the legitimate advertising industry. They are responsible for a lot of the things that you accept year after year as your problems. (1998: 108–109)

When we face complex, technical issues (as we do in multiple spheres of life), then these issues and technicalities need to be pondered, questioned, explained to lay citizens and debated. Due to the fact that politicians, increasingly guided by questionable 'experts' and technocrats, are subject to biases and flaws in thinking, we need people who can ask questions, and we need knowledgeable others who are willing to answer them (Nelson, 1978). In Europe, however, Morin (1999) observes a trend towards a process of democratic regression whereby citizens are dispossessed of major political decisions. For open dialogue to take place, we need democratic structures. A heterotopian perspective attends to regenerative countercurrents, therefore it pays tribute to the voices and social struggles of the rebellious minorities. The emergence of new political actors, such as *Barcelona en Comú* in Catalonia (see Russo & Scarnato, 2018), *Cooperation Jackson* (Akuno & Nangwaya, 2017) or *Poitiers Collectif* in France (Servigne *et al.*, 2020) reveals the hopeful and creative power of alternative voices.

The Structure of the Book

While cultural tourism cannot be disassociated from other sociopolitical processes, the work presented in this edited volume does not address the many entangled issues humanity will be confronted with beyond the tourism context. However, the scholars who have generously responded to our call to reflect on the likely future(s) of cultural tourism, have sought to stretch our understanding and initiate a conversation about what may lay ahead. In this volume, many authors look back, observe the present, ask pertinent questions about the future and provide some mind-provoking suggestions and scenarios for us to mull over. Before I introduce each chapter below, it is worth noting that there are many ways of telling and researching the future. The late John Urry (2016: 188) insightfully demonstrated that imagining social futures is complex, hazardous and uncertain because 'futures involve cascading interdependencies and wicked problems with multiple "causes" and "solutions"'. Because radical changes can happen from one day to the next, as the recent COVID-19 crisis has revealed, we (as editors) were reluctant to impose a specific time frame of what the contributing authors should understand as 'future'. The authors were, therefore, free to write about the future whether it be short, medium or long-term. Consequently, each chapter offers its own idiosyncratic perspective, with many chapters often addressing current issues with short-term solutions. Irrespective of time horizons, here, while one perspective flirts with dystopia (Chapter 10), many tend to be utopian. At the same time, a number of perspectives resonate with different shades of our heterotopian vision. The topics covered in the following chapters are diverse but could be broadly subsumed under four areas, namely: governance, consumption, technology and a general outlook.

The first part covers the topic of cultural tourism governance and includes four chapters. First, Prosper Wanner, Hakan Shearer Demir and Ivana Volić (Chapter 2) imagine a heterotopian future in which tourism and hospitality actors enact the concept of *community of hospitality* (communauté d'hospitalité). Community of hospitality refers to an autonomous, democratic body, which is strongly tied to the territory within which it operates, and which possesses agency to develop, control and manage cultural and natural resources. Communities of hospitality are based on municipalist governance principles, centered on belonging, dialogical exchange and responsibility for the common good. Wanner, Shearer Demir and Volić call cultural tourism actors to mobilise urban solidarities and to refuse the neoliberal logic of competitiveness, growth and profit-making. Then, in Chapter 3, Ella Björn and Monika Lüthje draw upon a posthumanist deconstruction of human-centred teaching and the notion of cultural sensitivity to envision how local cultures can be taught to tourists in new sustainable ways. Through interviews with Indigenous Sámi tourism entrepreneurs in Finnish Lapland, they reveal how social and material aspects as well as place and time are entangled. They reflect on how sociomaterial practices of teaching Sámi cultures might change cultural tourism in the future. In Chapter 4, in the context of the Dalmatian hinterland in Croatia, Lidija Petrić, Ante Mandić and Davorka Mikulić explore the role of cultural tourism villages in fostering sustainable and resilient rural communities. Their findings underscore the significance of cultural tourism villages as catalysts for local economic development, community empowerment and preservation of cultural heritage. As rural areas worldwide grapple with transformation, their study offers valuable insights into the potential of cultural tourism villages to shape resilient and vibrant rural communities. Chapter 5 – the fourth chapter in this part – is concerned with public participation in the conservation of World Heritage sites and highlights the need to pay attention not only to the construction of heritage values by authorised heritage discourse (AHD) but also to the public's understanding of these values. Here, Rouran Zhang, Weili Zhan, Ying Lyu and Da Kuang take the Longhushan (龙虎山) World Natural Heritage Site in Jiangxi Province as an example. They analyse tourists' perspectives of the site's heritage value by examining their online travelogues, which are then compared to the authoritative discourse and Chinese scholars' understanding of the site's value. Their findings reveal that tourists largely agree with the AHD's interpretation of the natural and cultural values of Longhushan and that these values are continuously evolving and enriching with time. Zhang and his colleagues also suggest that incorporating tourists' perspectives when preparing inscription texts would enrich the existing heritage value system and provide a more comprehensive and inclusive understanding of the site's heritage value. They argue that the Longhushan case resonates with the idea of creating heterotopias as spaces where alternative, resistant or emergent narratives can challenge or complement dominant ones.

The second part of this volume addresses consumption practices in future cultural tourism and includes three chapters. The first chapter, Chapter 6, written by Jelena Farkić, draws on feminist, new materialist theorising to discuss the possibilities for the development of slow cultural tourism. It does so by examining the emergent idea of heterotopian communities through making sense of the narratives, practices and experiences of four community members actively involved in tourism development in the central part of Serbia. The focus is maintained on the community's entangled relations with human and non-human actants, who have equal agency in co-designing, performing and making sense of slow cultural practices. Her study offers insights into the ways in which the community is being organised around slow, creative and authentic activities that hold immense cultural value. Farkić hopes to 'ecologise' our thinking by envisioning the future development of cultural tourism, transcending the boundaries of species, geographies and cultures. In Chapter 7, Dallen Timothy then follows in this part with a chapter that highlights several key trends that are worth considering in the future of cultural heritage-based tourism. Perhaps most prominently are deeper existential experiences that help people achieve self-actualisation and altruism in various forms. This chapter concludes with a utopian slant by suggesting that most modern changes and future directions are rooted in the self and individualisation of cultural tourism away from the mass consumption that has come to define modern cultural and heritage tourism. In Chapter 8, Marta Šveb Dragija and Daniela Jelinčić similarly point to a trend towards transformational museum experiences. Despite this trend, they note that it remains unclear how transformative experiences should be designed. To investigate this issue, using focus group conversations with cultural tourism practitioners in Croatia, they explore the future role of museums in promoting wellbeing, in terms of the design elements of the museum experience and in terms of the role of emotion in such experiences. Their chapter adopts a utopian vision of the future of cultural tourism.

The third part includes three chapters, which envision the use and influence of technology within the social world. This part opens with Chapter 9 in which Emanuele Mele imagines the influence of digital technologies on heritage experiences with the aim of producing a typology of cultural tourists from a utopian perspective. His conceptual analysis is informed by triangulating information from scientific articles, practitioners' viewpoints, as well as examples of pioneering initiatives in the fields of tourism, culture, and digital technologies. He describes future cultural tourists as cyber travellers, enhanced travellers, and creative travellers, whose experiences will be affected by the use of social media, artificial intelligence, augmented and virtual reality. In contrast to this, in Chapter 10, Daniel Wright envisages a dystopian future in which the use and implementation of artificial technology has led to greater controlled urban environments, inhibiting people's freedom for movement and travel. He notes, however, that past counterculture movements have proven to be an example of resilience towards top-down control. Here, the potential of countercultural movements in the future as a means of escaping urban digital prisons is considered. In line with our heterotopian perspective, Wright argues that it could be the travel behaviours of the marginalised that set the masses free. In Chapter 11, Ulrike Gretzel and Eva Sánchez-Amboage explore the foundations of the Metaverse and position it as a set of technologies that has already started to profoundly impact the ways in which culture can be experienced. Based on examples of existing museum offerings that apply Metaverse technology, these authors create a typology of immersive technology experiences that distinguishes between location-based and remote, and onsite, online and onlife experiences. This rather optimistic chapter concludes with a discussion of opportunities for cultural tourism in the Metaverse (from its potentially lower environmental and social impacts to its promise of heightened immersiveness, democratisation and greater accessibility of art experiences) as well as potential challenges (from its environmental cost to new digital divides and struggles of smaller, less well-funded museums to keep up with yet another technological wave).

The fourth and final part of the book includes three chapters and provides a general outlook on the future of cultural tourism. Because cultural and heritage tourism can play a critical role in improving both tourists' and residents' quality of life (QoL), in Chapter 12, Muzzo Uysal and Jiahui Wang examine the nexus that connects cultural and heritage tourism to QoL from a holistic perspective. They first trace the academic evolution of cultural and heritage tourism by employing a bibliometric approach and thematic analysis, which then helps them to articulate some possible ways how to foster QoL and sustainability of cultural and heritage tourism. In Chapter 13, Maurizio Scarciglia adopts an impressionist, literary style of writing to tell two utopian accounts of the future of cultural tourism destinations. To do so, Scarciglia recalls personal experiences and draws from current trends of spatial transformation to imagine two future urban environments: the resort city and the small provincial town. He also questions the value of existential authenticity as a concept through which to understand cultural tourism experiences. The readers are invited to make their own sense of his two fictional accounts. Chapter 14 concludes this book. Here, we offer a synthesis of the key contributions and some personal reflections on hopeful cultural tourism futures.

References

Akuno, K. and Nangwaya, A. (eds) (2017) *Jackson Rising: The Struggle for Economic Recovery and Black Self-determination in Jackson*. Daraja Press.

Arrieta Urtizberea, I., Hernandez Leon, E. and Andreu Tomas, A. (2016) Patrimonio local en un mundo global: Procesos de patrimonialización cultural en contextos locales de Andalucía y el País Vasco. *Memoria em Rede* 8 (14), 41–57.

Ateljević, I. (2020) Transforming the (tourism) world for good and (re)generating the potential new normal. *Tourism Geographies* 22 (3), 467–475.

Aubenas, F. and Benasayag, M. (2002) *Résister, c'est créer*. La découverte.

Benson, M.H. (2019) New materialism: An ontology for the Anthropocene. *Natural Resources Journal* 59 (2), 251–280.

Bostrom, N. (2009) The future of humanity. In J.K. Berg Olsen, E. Selinger and S. Riis (eds) *New Waves in Philosophy of Technology* (pp. 186–215). Palgrave McMillan.

Bostrom, N. (2019) The vulnerable world hypothesis. *Global Policy* 19 (4), 455–476.

Braidotti, R. (2019) A theoretical framework for the critical posthumanities. *Theory, Culture & Society* 36 (6), 31–61.

Case, A. and Deaton, A. (2020) *Deaths of Despair and the Future of Capitalism*. Princeton University Press.

Chomsky, N. and Pollin, R., with Polychroniou, C.J. (2020) *Climate Crisis and the Global Green New Deal: The Political Economy of Saving the Planet*. Verso Press.

Daniels, M.J. and Bowen, H.E. (2003) Feminist implications of anti-leisure in dystopian fiction. *Journal of Leisure Research* 35 (4), 423–440.

Debord, G. (1992) *La Societé du Spectacle*. Gallimard.

du Cros, H. and McKercher, B. (2015) *Cultural Tourism* (2nd edn). Routledge.

Farkić, J. and Taylor, S. (2019) Rethinking tourist wellbeing through the concept of slow adventure. *Sports* 7, Article 190.

Foucault, M. (1986) Of other spaces. *Diacritics* 16 (1), 22–27.

Herrington, G. with Wouters, R. (2023) Geopolitics beyond growth. *Green European Journal*, 3 May. See https://www.greeneuropeanjournal.eu/geopolitics-beyond-growth/ (accessed March 2024).

Higgins-Desbiolles, F., Carnicelli, S., Krolikowski, C., Wijesinghe, G. and Boluk, B. (2019) Degrowing tourism: Rethinking tourism. *Journal of Sustainable Tourism* 27 (12), 1926–1944.

Hollinshead, K. (2009) 'Tourism state' cultural production: The re-making of Nova Scotia. *Tourism Geographies* 11 (4), 526–545.

Houge Mackenzie, S. and Goodnow, J. (2021) Adventure in the age of COVID-19: Embracing microadventures and locavism in a post-pandemic world. *Leisure Sciences* 43 (1–2), 62–69.

International Court of Justice (ICJ) (2024, January 26) Order. Application of the Convention on the Prevention and Punishment of the Crime of Genocide in the Gaza Strip. See https://www.icj-cij.org/sites/default/files/case-related/192/192-20240126-ord-01-00-en.pdf (accessed April 2024).

Ivanov, S. (2023) What can we learn from Star Wars about the future of tourism? Absolutely everything! *Journal of Tourism Futures* 9 (2), 222–228.

Kumm, B.E. and Pate, J.A. (2024) 'This machine kills fascists': Music, joy, resistance. *Leisure Studies* 43 (3), 395–406. https://doi.org/10.1080/02614367.2023.2191982

Lafargue, P. (2023 [1880]) *Le droit à la paresse*. Librio.

Matteucci, X., Koens, K., Calvi, L. and Moretti, S. (2022a) Envisioning the future of cultural tourism. *Futures* 142, Article 103013.

Matteucci, X., Nawijn, J. and von Zumbusch, J. (2022b) A new materialist governance paradigm for tourism destinations. *Journal of Sustainable Tourism* 30 (1), 169–184.

Matteucci, X. and Smith, M.K. (2024) *The Creative Tourist: A Eudaimonic Perspective*. Emerald Publishing.

Matteucci, X. and Tiller, T.R. (2023) Package cycle tourists' relation to time and pace. *Tourism and Hospitality Research* 23 (3), 332–343.

Matteucci, X. and von Zumbusch, J. (2020) Theoretical framework for cultural tourism in urban and regional destinations. Deliverable D2.1 of the Horizon 2020 project SmartCulTour (GA number 870708). doi: 10.5281/zenodo.4785433

Monbiot, G. (2016) Neoliberalism – The ideology at the root of all our problems. *The Guardian*. See https://www.theguardian.com/books/2016/apr/15/neoliberalism-ideology-problem-george-monbiot (accessed December 2023).

Morin, E. (1999) *Seven Complex Lessons in Education for the Future*. UNESCO Publishing.
Mullis, K. (1998) *Dancing Naked in the Mind Field*. Vintage.
Nelson, T. (1978) Techno-politics. *Penthouse*, October, 53–54.
Perronne, C. (2021) *Décidément, ILS n'ont toujours rien compris!* Albin Michel.
Pies, R.W. (2020, October 13) Is the country experiencing a mental health pandemic? *Psychiatric Times* 37 (10). See https://www.psychiatrictimes.com/view/are-we-really-witnessing-mental-health-pandemic (accessed December 2023).
Pritchard, A., Morgan, N. and Ateljević, I. (2011) Hopeful tourism: A new transformative perspective. *Annals of Tourism Research* 38 (3), 941–963.
Raj, R. and Griffin, K. (2017) Introduction to conflicts, religion and culture in tourism. In R. Raj and K. Griffin (eds) *Conflicts, Religion and Culture in Tourism* (pp. 1–9). CABI.
Reijnders, S. (2009) Watching the detectives: Inside the guilty landscapes of inspector morse, baantjer and wallander. *European Journal of Communication* 24 (2), 165–181.
Richards, G. (2020) Designing creative places: The role of creative tourism. *Annals of Tourism Research* 85, Article 102922.
Richards, G. (2023, November 13) Prof Greg Richards on academic silos, localism, over-tourism, and modernity. *Tourism's Horizon*. See https://www.goodtourismblog.com/2023/11/professor-greg-richards/ (accessed December 2023).
Robinson, P., Lück, M. and Smith, S. (2020) *Tourism* (2nd edn). CABI.
Russo, A.P. and Scarnato, A. (2018) 'Barcelona in common': A new urban regime for the 21st-century tourist city? *Journal of Urban Affairs* 40 (4), 455–474.
Schaap, L. and Edwards, A. (2007) Participatory democracy. In M. Bevir (ed.) *Encyclopedia of Governance* (pp. 663–667). Sage.
Scherf, K. (2021) Creative tourism in smaller communities: Collaboration and cultural representation. In K. Scherf (ed.) *Creative Tourism in Smaller Communities: Place, Culture and Local Representation* (pp. 1–26). University of Calgary Press.
Seedhouse, D. (2020) *The Case for Democracy in the COVID-19 Pandemic*. Sage.
Servigne, P., Moncond'huy, L. and Fourreau, E. (2020) Inverser les imaginaires en écrivant un récit commun, mobilisateur et positif. *DARD/DARD* 2 (4), 90–109.
Sharpley, R. and Telfer, D.J. (2023) *Rethinking Tourism and Development*. Edward Elgar Publishing.
Sismondo, S. (2021) Epistemic corruption, the pharmaceutical industry, and the body of medical science. *Frontiers in Research Metrics and Analytics* 6, Article 614361.
Slow Movement (2024) Welcome to the slow movement. See https://www.slowmovement.com (accessed March 2024).
Smith, L. (2015) Intangible Heritage: A challenge to the authorised heritage discourse? *Revista d'Etnologia de Catalunya* 40, 133–142.
Spennemann, D.H.R. (2007) Of great apes and robots: Considering the future(s) of cultural heritage. *Futures* 39, 861–877.
Suleman, R. and Qayum, B. (2017) Consciousness in conflict. In R. Raj and K. Griffin (eds) *Conflicts, Religion and Culture in Tourism* (pp. 13–22). CABI.
Thompson, M. (2021) What's so new about New Municipalism? *Progress in Human Geography* 45 (2), 317–342.
Thoreau, H.D. (2017 [1849]) *Civil Disobedience*. Enhanced Media Publishing.
Toivonen, A. (2022) Space tourism – Science fiction becoming reality. In I. Yeoman, U. McMahon-Beattie and M. Sigala (eds) *Science Fiction, Disruption and Tourism* (pp. 56–70). Channel View Publications.
Tomassini, L. and Cavagnaro, E. (2022) Circular economy, circular regenerative processes, *agrowth* and placemaking for tourism future. *Journal of Tourism Futures* 8 (3), 342–345.
Tussyadiah, I. (2020) A review of research into automation in tourism: Launching the Annals of Tourism Research Curated Collection on Artificial Intelligence and Robotics in Tourism. *Annals of Tourism Research* 81, Article 102883.

Tzanelli, R. (2021) Post-viral tourism's antagonistic tourist imaginaries. *Journal of Tourism Futures* 7 (3), 377–389.

Urry, J. (2016) *What is the Future?* Polity Press.

Wright, D.W.M. (2023) The future past of travel: Adventure tourism supporting humans living on the edge of existence. *Journal of Tourism Futures* 9 (2), 151–167.

Yeoman, I. and McMahon-Beattie, U. (2020) Turning points in tourism's development: 1946–2095, a perspective article. *Tourism Review* 75 (1), 86–90.

Zorzin, N. (2015) Dystopian archaeologies: The implementation of the logic of capital in heritage management. *International Journal of Historical Archaeology* 19, 791–809.

Part 1
Governance

2 Communities of Hospitality as a Democratic Governance Model for Host–Guest Interactions

Prosper Wanner, Hakan Shearer Demir and Ivana Volić

Introduction

In the centre of Romania, Viscri, a small village of 430 inhabitants, has been resisting the dominant tourism system with its own framework and rules. Although Viscri receives 50,000 visitors each year (15,000 visitors stay overnight), the local residents have chosen to limit the season to eight months a year. The food served is entirely local, the souvenirs are handmade, the rooms are traditional and the lifestyle of its inhabitants (including animals) is not staged for tourists. The locals collectively make decisions to plan for future developments in the village, to address issues around social inclusion, income generation and to resolve most of the challenges the villagers are confronted with. Agriculture has remained the village's main economic activity. Accommodation providers do not compete to attract tourists and their number is limited. Travellers are welcomed by a community of villagers who have been living in harmony with the natural world. Hospitality rules and conventions are decided upon in an informal parliament and are enshrined in a social contract/agreement.

In the Provence region, Hôtel du Nord, a cooperative of residents in the northern districts of Marseille, has also been resisting the dominant forces of the tourism system. The members of this cooperative have joined forces to offer hospitality to all travellers, from tourists to caregivers; they tell their local stories and sell endogenous products. Their guiding principles are reflected in their motto, which says: 'we sell what we produce and we produce what we sell' (Hôtel du Nord, 2023), and which corresponds to an organic process of hospitality imbued with local solidarity and embedded within the northern district. The aforementioned two initiatives resonate with what Matteucci *et al.* (2022) have referred to as heterotopias, which, in the context of cultural tourism, they describe as pockets of resistance to mainstream neoliberal practices. Here, drawing from the real-life examples of Viscri and Hôtel du Nord, and building upon ongoing discussions in the field of cultural tourism, we argue that the value of heterotopia resides in enacting alternative political actions. In this regard, we propose the concept of *communities of hospitality*, as platforms able to facilitate a shift away from the dominant elitist approach to cultural tourism governance.

Tourism is often regarded as an activity that should receive widespread support and advocacy globally. This is because tourism is believed to bring benefits to both travellers and to the communities who host them (Gascón, 2019). While seeing tourism as a positive and mutually beneficial exchange may hold true in some contexts, questioning this assumption may seem to be a provocation. Adopting this simplistic viewpoint, we may endorse the idea that tourism development is inherently good as it generates a vast range of benefits to local communities. However, a more comprehensive and nuanced understanding of how the tourism industry operates reveals that tourism is often far from a purely benevolent activity. Furthermore, to use the terms 'hospitality' and 'tourism' interchangeably is problematic, because not all of those who receive hospitality services are affluent tourists. While this terminology may not make a difference to those within the tourism industry, it complicates the matter for migrants and refugees who do not enjoy access to the lofty services promised by the tourism industry.

A central concern in our discussion on cultural tourism here has to do with cultural rights for all visitors, irrespective of their backgrounds (whether these are tourists, refugees, migrants or asylum seekers). In this chapter, we therefore explore the nature of genuine mutual exchange, namely democratic and egalitarian relationships between guests and hosts. Considering that each locality is unique, and based on our understanding of cultural tourism, we advocate for self-governance processes, yet without prescribing any magic recipe. Instead, in this chapter, our objective is to stimulate dialogue around the concept of *community of hospitality* as a platform for enacting a shift from mere resistance to political action. To build our argument, we draw from the principles of *municipalism* and we address the following four questions:

- What do *communities of hospitality* understand by hospitality?
- Why do these communities need to define their own rules of hospitality?
- What does this tell us about the limits of tourism?
- How can *municipalism* contribute to public action?

We hope that the arguments presented below will stimulate critical discussions about alternative forms of cultural tourism governance.

Tourism as a Creator and Creation of Capitalism

Tourism can be understood as a product of capitalism since it is based on market principles and endless growth (Higgins-Desbiolles *et al.*, 2019). Gibson (2010: 521) describes tourism as a 'capitalist enterprise with an industrial and labor market structure'. As a result, tourism policies reflect the market logic – travel and hospitality are reduced to product(s) that need to be attractive to international tourists. The mainstream criteria used to measure the success of tourism destinations are an increase in tourists' overnight stays and expenditure, as well as an increase in accommodation and hospitality facilities (Liu & Liu, 2009; Volić, 2023). The idea of tourism as an enterprise comes from the United Nations World Tourism organisation (hereafter UNWTO). Globally, the values of growth and competitiveness have been automatically and uncritically applied to all levels of governance – state, regional and local (Bianchi & de Man, 2021). However, such an approach has been challenged as analysis by international institutions on the economic impact of tourism has decried the low community benefits of tourism, particularly in those countries with a low GDP. This finding thwarts the commonly touted benefits of tourism (Cousin & Réau, 2011; Caire, 2007; Duterme, 2018).

The problem with cultural tourism arises when places are transformed into products and some community members are driven by ideals of competitiveness, growth and profit-making. In such a context, a single narrative often prevails and serves the tourism industry and its clients. Local communities may feel compelled to present authentic performances to domestic and foreign visitors and, in turn, local cultures become commodified to fit marketed images and narratives. It is not uncommon that the tourism industry promotes fossilised cultures in order to stimulate demand and generate profit. Ultimately, tourism contributes to shape the identity of a place and it dwells in the narrative that speaks to the feelings and desires of tourists, while ignoring the concerns of local communities. Governmental bodies are increasingly expected to consult community members, as stakeholders in the grand scheme of business deals, yet being consulted does not necessarily mean making decisions. Feigning consultations tends to sideline multiple stories, including those of colonialism, social movements, the struggle for democracy and human rights. In such instances, tourism can be a self-destructive force to a community's pride, identity and dignity, as large companies gain more control over resources and local communities' economic benefits decrease (Chatterjee, 2010).

Commodified Encounters Versus Genuine Exchange

Hospitality, a basic exchange and interaction between human communities nurtured by mobility, has become a part of an industrial complex controlled by wealthy individuals and companies outside the community.

The concept of hospitality has been redefined to serve mainly profit-making purposes, removing the essence of mutual exchange from the context, as the term *hospitality* suggests in its original roots. The precious interaction between people (visitors and hosts in this case) and the stories about the significance of places are an opportunity for 'a moment of genuine exchange' that cannot be treated as merely a tourism transaction as defined by an industry. Viscri and Hôtel du Nord in Marseille serve as vivid and rare examples of such cases. However, the tourism industry prescribes parameters and expectations, thereby reducing this important moment of exchange to a superficial relationship to be consumed and perpetuating cultural extraction and appropriation. It is a form of coloniality reproduced in the tourism sector as 'colonialism is echoed in the imaginations of tourists, in the marketing of destinations and in the production of touristified landscapes' (Linehan *et al.*, 2020: 1).

Weakened by global economic instability and attracted by cultural imperialist inclinations, local communities are made to believe that they have limited opportunities to manage their resources and compete with large entities in the tourism sector. Liberalism creates and perpetuates a culture of fear by constantly reintroducing the spectre of danger and threat, leading to the expansion of control, constraints and coercion in order to normalise suppression (Mbembe, 2017). These dangers and threats include but are not limited to eroding democracies, financial instability and increasing precarity, increasing income gap, mass mobility toward urban settings, disappearing of traditional local practices, climate change and natural disasters. Consequently, inviting a well-established entity such as an international corporate chain to manage the business becomes more convenient – albeit to a degree which dehumanises this moment of exchange in the interest of profit-making.

Reclaiming the Core Meaning of the Word Hospitality

One of the challenges in studying the tourism system lies in the usage of the term 'hospitality'. Due to its Indo-European roots, which encompass various related words such as hostility, hostage, host, hotel, guest, hospital and hospitality, the word possesses multiple meanings and interpretations (Cinotti, 2011). This shared etymology encompasses the concepts of welcome, otherness and refuge. Given the diverse range of meanings associated with this term, it is essential to specify its intended context within research to prevent misunderstandings. The definition of hospitality varies depending on whether it is used in the social sciences or the management sciences. In management sciences, hospitality pertains to the tourism industry and the art of reception, primarily encompassing economic activities such as cafés, hotels and restaurants (Selwyn, 2000)[1]. In the social sciences, the term hospitality is employed in entirely different contexts, exploring topics such as identity, belonging, cultural interaction, displacement, migration, language, memory and cross-cultural blending (Gauvin & L'Hérault, 2004). While we acknowledge both definitions of hospitality, we would like to emphasise an integrated understanding of the term. We aim to explore the intersection where social and economic domains synergise and are enjoyed by all inhabitants at a community level thereby incorporating its political dimension.

Social scientists commonly distinguish between hospitality and tourist reception (Rabbiosi & Wanner, 2019). There is often an implication that tourist reception is more aligned with market terminology, while hospitality is associated with migrants and refugees. Consequently, the term hospitality carries a nuanced connotation that stems from its association with market-based and charity-based perspectives (Bauman, 1999; Boniface, 2014). In some cases, tourism development is even depicted as responsible for the death of hospitality as an essential form of socialisation (Scheou, 2010). Within the social sciences, hospitality is primarily understood as a face-to-face, interpersonal relationship where a host warmly welcomes a guest at their doorstep. This perspective considers hospitality as an individual virtue, emphasising the personal nature of the interaction (Agier, 2018). It portrays hospitality as a unilateral, isolated and punctual gesture of openness (Stavo-Debauge et al., 2018).

However, the prevailing influence of the profit-driven tourism industry on the concept of reception undermines the inherent meaning of hospitality, limiting it to the reception of migrants and refugees seen solely as acts of charity (Stavo-Debauge et al., 2018). The market's control over the term hospitality further perpetuates this erosion of meaning, transforming the social dimensions of hospitality into opportunities and commodities, driven by a liberal agenda (Chanial et al., 2019). The collaborative economy has played a significant role in the commodification of hospitality (Anspach, 2019). Online booking platforms (e.g. Booking.com and Airbnb) are good examples of commodified face-to-face encounters. It always implies monetary exchange as the basis for the encounter whether it involves consumer-to-consumer or business-to-business interactions.

This discourse is not new, and many authors, such as Diderot, D'Alembert and Gaubert, lamented in the 18th century that hospitality had become a business (Scheou, 2010). Montesquieu and Jaucourt asked themselves: 'Is the spirit of hospitality soluble in the spirit of commerce?' (Caillé et al., 2019: 28). For the anthropologist Cottereau, who focuses on ethno-accounting, Western thought tends to separate the economic domain, subject to monetary measurement, from other domains of social life, which is characterised by a plurality of values. Cottereau (2016) questions the classical definition of the economic object proposed by Vilfredo Pareto, who posited that economics studies the 'relationship of [wo]men to things' (2016: 11), while sociology focuses on the 'relationship between [wo]men' (2016: 11) in relation to things. These definitions, which distinguish between selfless hospitality and lucrative hospitality, and their opposition, become obsolete when we observe the hospitality practices of local communities on the ground.

This holds significant importance for our perspective, as we centre our focus on hospitality as a community virtue, emphasising its connection to community wellbeing rather than reducing it to a mere commodity. Our perspective is in line with the radical rethinking of the right to travel and hosting, proposed by Higgins-Desbiolles *et al.* (2019). In their opinion, tourism must be redesigned to 'acknowledge and prioritise the rights of local communities above the rights of tourists for holidays and the rights of tourism corporates to make profits' (2019: 1926). For us, it is also crucial to acknowledge that hospitality extends beyond being solely a tourism activity or an act of charity, which necessitates resources. Any attempt at approaching hospitality should involve a comprehensive understanding of its sociopolitical, cultural and economic implications on specific communities, without romanticising it.

The Need for a Shift (Towards Empowerment of Locals)

Our aim in this chapter is not to dwell on the corporate aspect of cultural tourism but to reveal the dialectical relationship between tourism and local communities and to examine whether the concept of hospitality can be reclaimed by local communities through a municipalist approach. Through such an approach, an alternative viewpoint of tourism and hospitality takes place in a political sphere where local communities do not merely 'participate in', but are the decision-makers, in a dignified process that considers the local resources, their use and management, and accessibility through fair profit-making and sharing by the local community. This creates a base for the development of a healthy local economic system based on fairness, equality and social entrepreneurship. It is also an opportunity to reimagine a community of hospitality with the perspective of receiving and treating people with respect and dignity, regardless of whether their mobility was instigated by choice (tourists, students, academics) or by force (migrants, refugees and asylum seekers). Indeed, this is an advocacy for political actions – an act of existence, identity and dignity of people from all walks of life involved in sociopolitical and socioeconomic processes. It is a shift from vulnerable and resilient communities dominated by business-as-usual practices to resisting and politically active communities acting for positive social change. Increased mobility, which has resulted in community regeneration, has a direct impact on community life, thus generating discussions on what constitutes our communities today, and what direction our communities should take in the future. Inevitably, mobility is not only a matter of choice but increasingly people are compelled to move by force due to armed conflicts, scarcity of resources, climate change and environmental disruptions.

Here it is important to emphasise that our reference to the local community is that of an autonomous body comprised of the people and the natural world, landscape, and all its characteristics; inclusive of all its inhabitants who have a connection to and are part of the everyday life of a place, regardless of their 'legal' status (Shearer Demir, 2021). Accordingly, our understanding of hospitality goes beyond its mere economic, profit-based and extractivist approach as defined by the industry and away from the ongoing reproduction of 'the cultural and economic practices of coloniality' (Alexeyeff & Taylor, 2016: 1). We refer to local community members, conscious of their heritage and cultural resources and valuing each member of their community, while engaging in a mutual exchange with the visitors. It is crucial to overcome the dominant focus on the absolute satisfaction of the tourist (increasingly carried out by migrant workers), providing services, entertainment, showcasing their local traditions and rituals as understood and desired by the tourists often at the expense of their context and significance. Such unconditional expectations, as long as it is paid for, perpetuate the coloniality of power (Quijano & Ennis, 2000)[2], which is constantly reproduced in the relationships of the neoliberal political and economic realm (Castro-Gómez & Restrepo, 2008).

People-oriented, community-based and participatory approaches to tourism play a crucial role in coping with, or adjusting to changing dynamics moving from a vulnerable to an empowering place in community life. With the change in demographics, a shift from a territorial/possessive connection to a relational connection to places, people and their stories can reshape the way we view hospitality. Relational connections that occur through meetings, collaboration and networking disrupt the current unidimensional capitalist governance and represent an embryonic attempt at radical change in how we practice hospitality (Matteucci *et al.*, 2022). It is particularly essential in increasingly urban settings and peripheries which are shaped by these relational connections and offer a more diverse and realistic representation of the places today.

When the 'Other' Becomes a Local

Diversity that comes with mobility (including refugees and migrants) is a crucial aspect to acknowledge, even if it may be avoided and invisible to mainstream culture. If visible, diversity often takes place through cultural appropriation and extraction, which is an asymmetric relationship between cultures and societies. The presence of minorities is often limited to the market, educational and cultural activities, while their visibility regarding sociopolitical life and democratic participation is absent through a lack of access or self-exclusion. This restricted relationship between dominant cultures and minorities and the marginalised is an invisible aspect of the tourism sector where a sizable migrant workforce is accepted as long as they provide services at the desired price and in a timely manner. The main narrative, authored by dominant groups, undermines the narrative of the other (Tuhiwai Smith, 2005), subtly relegating them to specific categories (as per their nation, class or caste, geographic origin, occupation, etc.), without allowing much space for their perspectives to be heard. This phenomenon, termed 'pseudospeciation' (Erikson, 1985), increasingly shapes the industry, solidifying commodified relationships between the served (tourists) and the servants (often a migrant group with 'cheap' labour), perpetuating the dehumanisation of those considered different. Outside of work hours, migrant workers are expected to be hidden from tourist eyes, cast away in their silenced peripheries. Therefore, 'the other' is deemed acceptable in the local context as long as they contribute to the labour force and provide certain cultural assets that can be extracted for tourism purposes. Multiple dimensions of their existence, daily life struggles and complexities in society, in general, are often not addressed in the process of commodification and commercial interests within the tourism sector.

Jeremy Rifkin (2001) warns that when the culture itself becomes absorbed into the economy, only commercial bonds will remain to hold society together. This raises the critical question of whether civilisation can survive when the commercial sphere becomes the primary arbiter of human life. Surviving and sustaining cohesive societies is not a natural process for communities; it requires a process of co-construction and adaptation that takes into account all aspects of community life including the social, economic, cultural and political dimensions. This is particularly important today, as we witness the ongoing regeneration of communities due to increased mobility. Consequently, the multiple identities and narratives that come with mobility need to have their place in community life for a constructive dialogue to occur, whether they are long-term residents or newcomers. Otherwise, we bear the risk of perpetuating tourism as an elitist notion for the few privileged and wealthy people. The tourism industry now develops and dominates the tourism narrative for local communities, which impacts the contexts and circumstances in which hospitality takes place. Such a position exhibits a paternalistic posture, as well as continuing coloniality of power, where the majority of the world works in the service of the privileged and powerful (Shearer Demir, 2023).

Community of Hospitality and Beyond

> L'hospitalité peut se définir comme le partage du 'chez soi', la mise en commun de l'acte et de l'art d'habiter. J'insiste sur le vocable habiter: c'est la façon d'occuper humainement la surface de la terre. C'est habiter ensemble (Ricoeur, 1997: 7).
> [Hospitality can be defined as the sharing of home, the sharing of the act and the art of living. I insist on the term inhabit because it is the way to occupy the surface of the earth humanly. It is to live together] (translated by the authors)

The concept *community of hospitality* relates to hospitality that is claimed and practised by a community of people of which the guest is a part. Stavo-Debauge and his colleagues (2018: 4) relate this kind of hospitality to 'habitability of the world in its forms'. The quality of a hospitable environment is one that invites one to stay, paying attention to conditions of the relationship beyond its sole financial terms. Such an environment facilitates mutualism and spontaneous encounters, leaving space for improvisation, without conditioning the genuine exchange between persons and the natural world. *A community of hospitality* is a group of people who practice the art of living together in an environment they seek to make and maintain hospitable. It can be composed of accommodation hosts, trusted third parties, guests, places, profit and non-profit practices and other forms of hospitality occurrences. Guests can stay with a host, they can be assisted by a trusted third party, be welcome in the public space, and be accompanied for a walk in the city by another person (Wanner, 2022).

The *community of hospitality* is a form of collective hospitality that is not proposed by the tourism system; one which essentially offers a relation from traveller to tourism professionals. A tourism system entails unilateral, isolated and temporary gestures which aim solely towards market transactions. The *community of hospitality* does not separate the social relationship from the economic one, just as it does not separate the hosts from one another or their living environment. The *community of hospitality* understands hospitality as the fruit of a common policy. It is a hospitality made up of rules, negotiations and exchanges which can be for money or not. The combination of all these elements creates an essence of hospitality as an art of living (and hosting) together. The concept of *community of hospitality* considers the plurality of people welcomed and it embraces a diversity of forms of hospitality. This concept also promotes the creation of political spaces for dialogue on hospitality as the right to the city and the right to free movement. It refers explicitly to the notion of heritage community as defined in the Council of Europe Framework Convention on the Value of Cultural Heritage to Society known as the Faro Convention[3]. The Faro Convention defines *heritage communities* as

> self-organized and self-managed groups of individuals who are interested in the progressive social transformation of relationships between people, places and stories, with an inclusive approach based on an improved definition of heritage. [...] They promote direct democratic engagement, diversity and sustainable local development focused on heritage and the pursuit of economic and social conditions conducive to the life and well-being of diverse communities.

The concept of *community of hospitality* refers to those groups of people who organise themselves, formally or informally, to jointly create an offer for guests. Examples of community of hospitality principles that we are presenting here are the ones from Marseille (France) and Viscri (Romania). In Marseille, a local community of hospitality is called Cooperative Hôtel du Nord. Hosts of this community organise themselves collectively and cooperate with each other so that they can offer hospitality to their guests (Wanner, 2017). At Hôtel du Nord, a guest is not only a visitor coming for leisure; a guest is anyone who needs hospitality at a certain moment. For example, if a family member of someone hospitalised in the Hôpital Nord de Marseille needs accommodation, the hospital staff connects him/her with a referee in the Cooperative Hôtel du Nord. This referee directs a family member to a host that can accommodate him/her, and also to people who could facilitate his/her stay.

Hospitality in Hôtel du Nord takes different shapes and forms; it could be free or paid, or even distributed over several hosts, and often all at once. Service in Hôtel du Nord is not based simply on monetary exchange. In the core of hospitality at Hôtel du Nord is a democratic decision-making process about what kind of offer should be provided, based on the capacities of hosts. This means that the community takes into account the financial and psychological status of hosts. This contrasts with what Derrida (1997) calls 'pure reception' which does not consider one's identity, desires, rules, language, his/her capacity to work, integration and adaptation. The cooperative negotiates all these elements and decides collectively which services will be charged and how much. Similar principles apply to guests; it is important to know his/her status to create a customised service. The rules of the cooperative are somewhere in between Derida's free services as an art of generosity, and hospitality practised within the tourism industry that relies strictly on monetary exchange.

The village of Viscri is another example of a community of hospitality. The main tool for deliberation and decision-making about tourism is a social contract/agreement. A social contract/agreement is the space where the inhabitants, engaged or not in a tourist activity, collectively set a certain number of rules for managing tourism development and navigating the growing tensions generated by tourism success. Such a form of agreement implies the reversibility of decisions created by the community members; all decisions are prone to change in order to adapt to current needs and challenges. Similar to Hôtel du Nord, income generation and social inclusion issues are identified and considered collectively. Local cultures in Viscri are presented in every aspect of the hospitality – the houses for accommodation are restored according to the highest architectural standards for the vernacular Saxon architecture. In addition, every aspect of the offer is based on the cultural practices of residents (e.g. food is grown and prepared locally, handicraft products are from the village, stories being told are stories of residents). A further cultural layer in Viscri manifests as a culture of deliberation and participative decision-making for the benefit of the community. In our opinion, the latter cultural layer is what we see important when envisioning future cultural offers for visitors – the open and enabling environments for discussion and determination what culture is for every resident and how to continue (or not) with presenting that culture to visitors.

Cooperation in the framework of public action in Marseille and Viscri, as pockets of resistance (Matteucci et al., 2022), takes place in an informal setting and is the result of the goodwill of local institutions and political actors. The Faro Convention is a framework convention, which creates no rights and could be only ratified by States. The social contract/agreement in Viscri and the cooperative of inhabitants in Marseille are unique frameworks invented by the local communities. A municipalist approach in Marseille, as in Viscri, would create a space for further exploration and application of the principles of the Convention at the local level. Such an approach would reduce the dependence of local communities on local institutions and feed into dual power as political players while strengthening hospitality communities' role as essential actors in decision-making.

Putting Communities of Hospitality in Action Through Municipalism

Working towards a tourism future and reclaiming the essence of hospitality is a practice of direct democracy; a process of co-construction of communities, an effort toward a 'community of equals' as described by Jacques Ranciere (Baiocchi & Connor, 2013)[4], and equitable societies. Twenty-first century alternatives in governance critically question and reject pre-existing patriarchal and hierarchical structures. Exploring a local governance model that works towards a more equitable society seeks an institutional form based on self-management that supports local inhabitants to engage through direct democracy. It involves all members of society participating directly, equally and consciously in a collective process of shaping institutions.

A social theorist and political philosopher Murray Bookchin emphasises that reclaiming traditional values of mutual aid and complementarity within a 'social ecosystem' that transcends mere economic exchange can model a more holistic vision of the human community (Tokar, 2010). According to this view, a 'moral economy' and direct democracy at local levels serve as a kind of school for community co-construction with all inhabitants, fostering social power deriving from the community. A local municipalist structure with the municipalisation of local resources may create an environment where the community's agency can bring about change. Cognisant of the challenging nature of such an approach, along with a well-established industrial complex, all efforts on alternatives in tourism must build on one another. These efforts will need to go beyond the perceived image of the exotic places and people, and achieve gravitas in order to challenge a system of privilege that excludes multiple narratives, raises unsustainable expectations and perpetuates commodification of culture and heritage in relation to community lives. Communities can claim the commons and actualise their potential in free municipalities that are rationally and discursively constituted and institutionalised (Bookchin, 2015).

Municipalism, a term interchangeable with communalism, is the all-encompassing term given to a comprehensive theory and practice that seeks to reconstruct society along ecological lines[5]. It is based on the essential premise that all environmental problems are rooted in social problems (Amargi & Amargi, 2015). It focuses on reclaiming the public sphere to exercise authentic citizenship, freedom and responsibility and addresses people's aspiration for a quality of life beyond mere survival. It is a form of local governance that is primarily concerned with everyday problems and considers all residents capable of making their own decisions about their neighbourhoods, towns and cities instead of delegating to others. Municipalism, as a strategy for challenging the neoliberal political and economic order, allows local communities to express their collective decisions, cooperating to create a balance in power (Tarinski & Olney, 2019). Accordingly, change will require thinking and acting outside of the framework of the current system, beyond a centralised power. In this regard, reclaiming the commons and deciding on the utilisation and management of public space and public life is at the centre of municipalist action. The aim is to build grassroots power with local people and the capacity to expand itself. In this politics of proximity, face-to-face interaction is crucial for redefining and redesigning relations as an organic platform for a community of hospitality to operate. Starting from everyday life practices through popular assemblies, the implementation of direct democracy is an empowering act, shaping the present and the future of communities.

Direct representation of all present, and the principles of self-management, should be considered essential elements of educational, work and social life toward the co-construction of a new narrative of communities of hospitality in a network that offers alternative ways of practising hospitality. Although diverse cultural settings may require different strategies in the realisation of direct democracy, the struggle against structural injustices requires local community members to be a part of everyday life with the consciousness of hierarchical and patriarchal relations across cultures and geographies. The specificity of such structures must be left to the local communities themselves, using the principles of direct democracy, such as assemblies and networks, to organise, envision and put alternatives into practice. Recurring questions to be periodically posed are:

> Tourism and hospitality for whom, in the expense of what, and what are we reproducing with the host-guest relations to ensure hospitality is not a static or dogmatic notion of a tourism industry only serving the few but remains as a valuable moment of genuine exchange between people?

The experiences in Viscri and Marseille have shown that while there are set objectives and rules, the specific content and path are developed within the process in each unique case. Therefore, an inductive approach, where practice informs theory, might be more appropriate in order to create an enabling environment for pockets of resistance (Matteucci *et al*., 2022) to explore future possibilities in cultural tourism, without preconditioning the outcomes.

Notes

(1) It has been used since 1975 in the British Library catalog as a keyword for books that deal mainly with the economic activities of cafés, hotels and restaurants (Selwyn, 2000).
(2) The 'coloniality of power' is an expression coined by Anibal Quijano to name the structures of power, control and hegemony that have emerged during the modern era, the era of colonialism, which stretches from the conquest of the Americas to the present.
(3) https://www.coe.int/en/web/culture-and-heritage/faro-convention
(4) Jacques Rancière's definition of creating a community of equals, emphasises that equality is not a goal to be reached, but requires constant effort. Equality does not have a fixed arrival point where it becomes a 'substantial form as a social institution' but is a collective action by all inhabitants to adjust themselves to dynamic elements and changes in communities.
(5) Municipalism entails the main principles of popular assemblies, dual power, feminisation of politics, social ecology and confederalism.

References

Agier, M. (2018) *L'étranger qui vient: Repenser l'hospitalité*. Seuil.
Alexeyeff, K. and Taylor, J. (2016) *Touring Pacific Cultures*. ANU Press.
Amargi, S. and Amargi, M. (2015, May 26) Communalism: A Liberatory Alternative. See http://new-compass.net/articles/communalism-liberatory-alternative (accessed June 2023).
Anspach, M.R. (2019) Économie du partage ou capitalisme sauvage? Sur la marchandisation de l'hospitalité. *Revue du MAUSS* 53 (1), 55–64.
Baiocchi, G. and Connor, B.T. (2013) Politics as interruption: Rancière's community of equals and governmentality. *Thesis Eleven* 117 (1), 89–100.
Bauman, Z. (1999) *Le coût humain de la mondialisation*. Hachette Littératures.
Bianchi, R.V. and de Man, F. (2021) Tourism, inclusive growth and decent work: A political economy critique. *Journal of Sustainable Tourism* 29 (2–3), 352–370.
Boniface, P. (2014) *50 idées reçues sur l'état du monde* (4e éd.). Armand Colin.
Bookchin, M. (2015) *The Next Revolution: Popular Assemblies and the Promise of Direct Democracy*. Verso.
Caillé, A., Chanial, P., Gauthier, F. and Robertson, F. (2019) Le don d'hospitalité. Quand recevoir, c'est donner. *Revue du MAUSS* 53 (1), 5–26.
Caire, G. (2007) Tourisme solidaire, capacités et développement socialement durable. *Marché et organisations* 1 (3), 89–115.
Castro-Gómez, S. and Restrepo, E. (2008) *Genealogías de la colombianidad: Formaciones discursivas y tecnologías de gobierno en los siglos XIX y XX*. Pontificia Universidad Javeriana.
Chatterjee, A. (2010) Tourism Industrial Complex. Year of No Flying. See www.yearofnoflying.com/2010/01/tourism-industrial-complex.html (accessed June 2023).

Cinotti, Y. (2011) Hospitalité touristique: Conceptualisation et études de l'hospitalité des destinations et des maisons d'hôtes. Unpublished PhD thesis, Université de Perpignan Via Domitia.

Cottereau, A. (2016) Ne pas confondre la mesure et l'évaluation: Aspects de l'ethnocomptabilité. *Revue des Politiques Sociales et Familiales* 123 (1), 11–26.

Cousin, S. and Réau, B. (2011) L'avènement du tourisme de masse. *Les Grands Dossiers des Sciences Humaines* 3 (22), 14–14.

Derrida, J. and Dufourmantelle, A. (1997) *De l'hospitalité*. Calmann-Lévy.

Duterme, B. (2018) Tourisme Nord-Sud: Le marché des illusions. *Alternatives Sud* 25 (3), 7–27.

Erikson, E. (1985) Pseudospeciation in the nuclear age. *Political Psychology* 6 (2), 213–217.

Gascón, J. (2019) Tourism as a right: A 'frivolous claim' against degrowth? *Journal of Sustainable Tourism* 27 (12), 1825–1838.

Gauvin, L. and L'Hérault, P. (2004) Introduction. In L. Gauvin, P. L'Hérault and A. Montandon (eds) *Le dire de l'hospitalité* (pp. 7–16). Clermont-Ferrand: Presses Universitaires Blaise Pascal.

Gibson, C. (2010) Geographies of tourism: (Un)ethical encounters. *Progress in Human Geography* 34 (4), 521–527.

Higgins-Desbiolles, F., Carnicelli, S., Krolikowski, C., Wijesinghe, G. and Boluk, K. (2019) Degrowing tourism: Rethinking tourism. *Journal of Sustainable Tourism* 27 (12), 1926–1944.

Hôtel du Nord (2023) La Coopérative d'habitants. See www.hoteldunord.coop/la-cooperative-dhabitants/ (accessed June 2023).

Linehan, D., Clark, I.D. and Xie, P.F. (2020) Introduction. In D. Linehan, I. D. Clark and P.F. Xie (eds) *Colonialism, Tourism and Place: Global Transformations in Tourist Destinations* (pp. 1–11). Edward Elgar Publishing.

Liu, A. and Liu, H.-h.J. (2009) Government approaches to tourism: An international inquiry. *International Journal of Tourism Policy* 2 (3), 221–238.

Matteucci, X., Koens, K., Calvi, L. and Moretti, S. (2022) Envisioning the futures of cultural tourism. *Futures* 142, Article 103013.

Mbembe, A. (2017) Difference and self-determination. *e-flux Journal* 80, 1–12.

Quijano, A. and Ennis, M. (2000) Coloniality of power, Eurocentrism, and Latin America. *Nepantla: Views from South* 1 (3), 533–580.

Rabbiosi, C. and Wanner, P. (2019) Dal 'diritto Alla città' Al 'diritto Alla mobilità'. Spunti Per Una Critica Socio-Spaziale Della Definizione Di 'turista'. *Scritture Migranti* 13, 129–153.

Ricoeur, P. (1997) Etranger, moi-même. Conférence donnée au cours de la session 1997 des Semaines sociales de France, 'l'immigration, défis et richesses' (pp. 1–11). See www.ssf-fr.org/articles/54123-etranger-moi-meme (accessed June 2023).

Rifkin, J. (2001) *The Age of Access*. TarcherPerigee.

Scheou, B. (2010) Le retour de l'hospitalité, pratiques subversives ou expression d'une conformité postmoderne? *Revue L'autre voie* 6, 1–22.

Selwyn, T. (2000) An anthropology of hospitality. In C. Lashley and A. Morrison (eds) *In Search of Hospitality: Theoretical Perspectives and Debates* (pp. 18–37). Butterworth-Heinemann.

Shearer Demir, H. (2021) *ST21 MooC – implement ST21 in 10 steps – Culture and Cultural Heritage*. Council of Europe. See www.coe.int/en/web/culture-and-heritage/strategy-21 (accessed June 2023).

Shearer Demir, H. (2023) *Displacement Governance and the Illusion of Integration*. Springer.

Stavo-Debauge, J., Deleixhe, M. and Carlier, L. (2018) Hospitalités. L'urgence politique et l'appauvrissement des concepts. *Sociologies*, 1–29. https://doi.org/10.4000/sociologies.6785

Tarinski, Y. and Olney, E. (2019) *Direct Democracy and the Passion for Political Participation*. ROAR Magazine. See https://roarmag.org/essays/direct-democracy-yavor-tarinski/ (accessed June 2023).

Tokar, B. (2010) Can we buy our way to an ecological society? *Communalism: A Social Ecology Journal 2* (Spring/Summer), 10–12.

Tuhiwai Smith, L. (2005) *Decolonizing Methodologies: Research and Indigenous Peoples*. Zed Books.

Volić, I. (2023) Tourism policy values in Serbia – From equity to competition. *Tourism Planning & Development* 20 (5), 901–918.

Wanner, P. (2017) De l'exercice du droit au patrimoine culturel. In S. Pinton and L. Zagato (eds) *Cultural Heritage. Scenarios 2015–2017* (pp. 53–68). Edizioni Ca' Foscari.

Wanner, P. (2022) Tourisme social, économie collaborative et droits culturels: Ethnographie d'une coopération complexe. Unpublished PhD thesis. Université Paris-Nanterre.

3 Envisioning Posthumanist Cultural Tourism: Indigenous Sociomaterial Practices of Teaching Tourists about Local Cultures

Ella Björn and Monika Lüthje

Introduction

In this chapter, we discuss sociomateriality in teaching and learning about local cultures in tourism. We envision new posthumanist, sustainable and culturally sensitive ways to teach Sámi and other local cultures to tourists in Finnish Lapland. The Sámi are an Indigenous people living in *Sápmi* (Sámiland), which covers the northern part of Finland, northern and central parts of Sweden and Norway and the Kola Peninsula in Russia. The appropriation and misrepresentation of Sámi cultures by non-Sámi tourism companies have historically caused many problems in Finnish Lapland. Because much incorrect information circulates about the Sámi, providing correct cultural information to visitors is considered by the Sámi Parliament (2018) to be important for safeguarding Sámi cultural heritage and offering sustainable tourism solutions.

Although the situation has improved, fake Sámi culture (e.g. imitations of Sámi handicrafts and exotic rituals invented by tourism companies) is still offered to tourists, who do not receive enough correct information about authentic Sámi cultures and products (Kugapi *et al.*, 2020; Saari *et al.*, 2020). Today, certain niche types of tourists particularly seek to experience authentic local cultures when travelling (see Matteucci *et al.*, 2022a), and they are interested in Indigenous cultures (Kramvig & Førde, 2020). Introducing new culturally and socially innovative tourism products may help to develop more year-round tourism and improve the well-being of the people in *Sápmi* where tourism has been seasonal and nature-based. Culture-based products can be offered to tourists regardless of the season and enhancing the collaboration between the creative and tourist industries may help to address seasonality problems while bringing sources of livelihood to local handicraft makers and artists.

Posthumanism aims to break down Western humanist dichotomic categorisations like nature and culture (Cohen, 2019) – upon which the categorisation of nature-based versus cultural tourism is also built. In Sámi cultures, human beings, nature and culture are entangled. Connection to the land is, to the Sámi, a social tie that should be respected (Kuokkanen, 2007). The close relationships of Indigenous cultures with nature are often characterised as sustainable and harmonious. Although the Sámi nature relationship is considered close, not all Sámi have a nature-bound way of life. Their nature relationship is practical, local and experientially anchored in certain places and environments (Guttorm, 2021; Valkonen & Valkonen, 2014).

Posthumanist deconstruction of human-centred knowing involves the agency of the non-human, social and material aspects and attributes of place and time in the practices of teaching tourists about Sámi cultures. Understanding the role of sociomaterial agents helps build memorable experiences through comprehensive learning about different cultures while travelling. Place-based approaches in teaching tourists about culture in natural environments is one way to address the agency of nature and the role of sociomateriality in tourist learning processes and to bring new methods to cultural tourism planning. Acknowledging the role of sociomateriality in teaching brings new knowledge of the roles of different agents in the cultural tourism experience both for visitors and locals. As Ingold (2011) states, the world is not static but rather a *meshwork* of relations of humans and other beings inhabiting the world. Movements of beings along the lines of the meshwork produce place (Ingold, 2011), which means that the place identity is formed from encounters of locals and tourists and sociomaterial elements. Storytelling as a teaching practice is one way to point out these relations. Sociomaterial teaching can enhance tourist understanding that the place is formed by different encounters and relations, and the culture is not statically separated from place and time.

Culturally sensitive tourism, combined with sociomaterial knowledge sharing, can offer ways to preserve local cultural heritage anchored to certain places and local practices. Cultural sensitivity is an open, situational and relational orientation towards otherness and the other that offers a continuous possibility to question, reflect upon and change one's own being, knowing and values in relation to the other. It is a disposition that both tourists and locals can develop by reflecting on their prejudices, norms and values (Viken *et al.*, 2021). Recently, cultural sensitivity has been conceptualised in particular in the context of Indigenous tourism (Hurst *et al.*, 2021; Viken *et al.*, 2021), but it is a useful concept in all cultural tourism (Marques & Oliveira, 2023).

The future of the whole planet is threatened by the Western way of life and neoliberal growth-driven economic systems where nature, other people and their cultures are exploited as resources for the comfort of a few. Posthumanism is needed to focus attention and action on the well-being of the planet and all its living creatures (Guia & Jamal, 2023; Matteucci *et al.*, 2022b). In an increasingly global and multicultural world, cultural sensitivity is also highly relevant. It is needed for peaceful co-existence of humans belonging to different cultures. On one hand, there is globally a growing interest in and appreciation of Indigenous peoples and their cultures, partly because of their relation to nature, nature-related knowledge and a more sustainable way of life (Junka-Aikio, 2019, 2022). On the other hand, they suffer worldwide even today from colonialism and racism, which are also present in tourism (Holder *et al.*, 2023; Kugapi *et al.*, 2020). Hence, there is a need to carry out tourism differently. Sociomaterial understanding can help bring out the roles of the non-human in tourism products and services, build respect towards other cultures and living beings and foster a vision of a more sustainable future in the cultural tourism field. Teaching about culture through sociomaterial approaches in tourism can help move towards these goals.

This chapter focuses on sociomaterial practices of teaching about Sámi cultures to tourists on the basis of interviews of Sámi tourism entrepreneurs in Finnish Lapland. The interviews were conducted for the Culturally Sensitive Tourism in the Arctic (ARCTISEN) project (see Kugapi *et al.*, 2020). They included entrepreneurs' views on teaching tourists about local cultures and offer rich insights into the role of sociomateriality in the local tourism practices. In the following, we combine these insights with posthumanist and culturally sensitive perspectives.

Theoretical Background

Relational approaches like posthumanism, sociomateriality and cultural sensitivity 'emphasise the primacy of relations and dependencies' (Birhane, 2021: 4) between 'the subject(s) and object(s) that constitute each other and co-emerge from the relationship they establish and develop' (Cerratto Pargman, 2023: 42). This is an interesting point of departure for developing Indigenous cultural tourism in which Sámi tourism entrepreneurs and tourists are in a relationship with each other and a variety of other human and non-human subjects and objects, such as reindeer and reindeer herders, snow, snowmobiles, clothing and other equipment, handmade or industrial. In this fabric of sociomaterial relationships, both human and non-human actors are the result of the relationships they have with each other (Decuypere & Simons, 2016). Posthumanist educational theory sees all actors in these relationships – humans and non-humans – as teachers. In posthumanist theory, teaching is understood as 'the condition affording learning – or simply causing change in the learner' (Kouppanou, 2022: 773–774).

Viken *et al.* (2021) offered a theoretical model for analysing various ways in which human actors relate to cultural differences in tourism encounters and product development. The model was adapted from Bennett's (1986) model of intercultural sensitivity. Viken *et al.* (2021) defined cultural sensitivity as a subjective orientation towards otherness. They did not see it as a competence or skill but, rather, as a disposition that everyone can develop by reflecting on their own prejudices, norms and values. As such, cultural sensitivity is an ethno-relative orientation. It is defined as a relationship between the self and the other and is a situational, not permanent, quality. In contrast, cultural insensitivity is a self-centred, ethnocentric orientation that detaches the self from others. It is an orientation in which 'cultural differences are categorised, perceived as inferior to one's own, or reduced to sameness' (Viken *et al.*, 2021: 4; see also Bennett, 1986). In the model from Viken *et al.* (2021), cultural insensitivity is exemplified by assimilation, stereotyping and cultural appropriation, while cultural sensitivity consists of recognition, respect and reciprocity.

For Viken *et al.* (2021), the recognition of the other and otherness are the starting point of cultural sensitivity. They offer a continuous possibility of questioning, reflecting, and changing your own being, knowing and values in relation to the other (Viken *et al.*, 2021). However, Viken *et al.* (2021) were not explicit about their definition of 'recognition'. In recognition theories, the discussion of recognition can be divided into two perspectives: reciprocity and regard. In the first perspective, recognition is an interactive relationship between two humans, while in the second perspective, beings or entities other than humans, such as animals, culture or nature, can also be recognised (Laitinen, 2010; see also Kortetmäki, 2020; Lüthje, 2023).

From the perspective of reciprocity, recognition is a relationship in which another person is considered a significant interaction partner (Hirvonen, 2020). The relationship is mutual; to feel recognised, a person must give recognition to their recogniser. According to this theory, 'there always needs to be two-way recognition for even one-way recognition to take place' (Laitinen, 2010: 319). For example, taking another person's opinions seriously is a form of recognition, while underrating them means there is no recognition of the other (Hirvonen, 2020; Kortetmäki, 2020). Getting recognition is important because it is a precondition of human identity formation, agency and freedom (Laitinen, 2010).

In recognition theories, respect is a form of reciprocal recognition. Reference is often made to Honneth (1992), in whose theory respect means seeing all humans as being similar in their essential characteristics. Thus, all humans are equal discussion partners with the same rights and responsibilities in society or the world (Hirvonen, 2020; Särkelä,

2020). Here, 'rights' normally means human rights or other fundamental rights belonging to everyone (Turtiainen, 2020). Respect also means exercising one's rights in a manner that does not prevent others from exercising their rights (Särkelä, 2020; United Nations, 1948), which is an essential point of view when discussing the rights of local people, tourists and tourism companies in tourist destinations (Higgins-Desbiolles *et al.*, 2019).

Recognition can also be seen as respect for the particularity of a culture, where the particular characteristics of each culture and cultural group are taken into account equally in terms of equal worth being afforded to all cultures and groups (Kortetmäki, 2020; see also Assmann, 2013; Bennett, 1986). Here, recognition can also be understood as adequate regard, where the quality of the regard depends on the particular culture's needs or what is appropriate in relation to the culture's particular characteristics. The adequacy of the regard is evaluated in comparison to the best available regard (Laitinen, 2010). In the case of the Sámi, adequate regard may consist of, for instance, not disturbing reindeer, who roam free in nature, or not littering in Sámi sacred sites in nature. Because nature is an important part of Sámi cultures, cultural sensitivity also entails recognising the particular characteristics of local nature in tourism in *Sápmi* (see Kortetmäki, 2020), which is also essential from the point of view of sustainability and posthumanism.

In summary, cultural sensitivity consists of both recognition of cultural differences and giving equal worth to all human beings regardless of their cultures (see Bennett, 1986). Both should be affirmed in encounters between tourists and locals on both sides (Assmann, 2013). In a similar vein, Viken *et al.* (2021: 9) suggested that cultural sensitivity thrives 'in encounters that enable reciprocal exchange between hosts and guests through sharing and receiving, (un)learning and teaching'. Ideally, culturally sensitive tourism fosters reciprocity when tourists, tourism entrepreneurs, and other locals care for each other's needs and wellbeing, and this results in shared affective or ethical values, such as friendship, respect, compassion, trust and mutual understanding (Viken *et al.*, 2021; see Sabourin, 2013; Walsh-Dilley, 2017).

In addition to relationality, cultural sensitivity resembles posthumanism in its post-dualism: understanding that human experience is pluralistic, rather than seeing it in generalised and universal terms or dualistic categories, particularly if the categories are hierarchical (see Guia & Jamal, 2023). Hence, cultural sensitivity means recognising inter- and intracultural diversity. Sámi cultures are also diverse. For example, in Finland, three different Sámi languages – Inari, Skolt and North Sámi – are spoken and their speakers have differing historical backgrounds and cultures. Nowadays, many Sámi also live outside *Sápmi*. Sámi researchers emphasise that there is no one Sámi voice that all Sámi share but a polyphony of voices (Junka-Aikio, 2019; Magga, 2018).

The posthumanist paradigm allows us to examine knowledge creation in tourism through different practice-based theories leaning towards relationships, networks, materialities and an emergent reality. Practice-based studies acknowledge the social, historical and structural context on the basis of knowledge creation (Corradi *et al.*, 2010). Knowledge can be considered collective, social and collected (Gherardi, 2011), including tacit knowledge (Corradi *et al.*, 2010). Lave and Wenger's (1991) constitution of *communities of practice* acknowledges newcomers and practitioners, like tourists, in acquiring skills, absorbing culture-specific values, and developing identities while incrementally becoming a part of a community of practice.

Law and Urry (2004) described reality as being relational and produced in the interaction of the material and the social. Actor-network theory (ANT) focuses on how social arrangements, orderings, and materialities are accomplished and how tourism *works* instead of *what* tourism is (van der Duim *et al.*, 2013). ANT describes the different orderings through which our world emerges. Knowledge and agency are performed, reproduced and distributed through social networks by always taking material forms (Law, 1992; Nicolini, 2012; van der Duim *et al.*, 2013). The sociomaterial perspective sees all things – human, non-human, hybrid, knowledge and systems – as a sequence of connections and activity. Things come into existence in webs of relationships, and the practices through which boundaries come into being are highlighted (Fenwick *et al.*, 2015). There is a deep connection between action and experience, which showcases circularity in knowledge (Escobar, 2018). Maturana and Varela (1987: 26) summarise it as, 'all doing is knowing, and all knowing is doing'. Knowledge accumulates through encounters with others, but also through stories heard. 'To tell a story is to relate, in narrative, the occurrences of the past, bringing them to life in the vivid present of listeners as if they were going on here and now' (Ingold, 2011: 161).

People create places through embodied practices (Pink, 2008). Senses form an important part of embodied experiences while teaching and learning about local cultures and places. Knowledge rooted in practice considers not only people's interrelated minds but also corporeality leading to sensible knowledge creation, in relation to non-human elements (Strati, 2007). Embodied practices, together with senses and material, non-human elements, offer new sustainable ways to co-create and build an understanding of the world. Over time, Sámi cultures have, like other Indigenous cultures, been shaped by and adjusted to place, nature and climate, which can be seen in the material and non-material elements of these cultures. Although the culture–nature relationship produces Sáminess, it needs to be considered that there is a great ecological diversity of Sámi cultures and that cultures change over time, which means that the human–nature relationship is a highly complex and multidimensional issue (Valkonen & Valkonen, 2014). Nature relations are more local, discursive and practical than general, revealing more about the lives of people in a certain area (Valkonen & Valkonen, 2014), which needs to be considered when creating an understanding of the sociomaterial aspects of teaching and co-creating knowledge in tourism.

Research Methods

Our research focused on the *practices* of teaching and learning about local cultures and places that are constructed through social and material relations. The practices consist of knowledge-building and sharing, utilising embodied activities, senses, materials (non-human) and social connections. The research material included transcribed interviews with tourism entrepreneurs in Finnish Lapland conducted for the ARCTISEN project in 2019. The project offered insights into local values and cultural assets that formed a basis for planning new cultural tourism products. Through a practice-based lens, we describe how tourists can be taught about local cultures by utilising materiality, doing, and place-based and embodied approaches.

The ARCTISEN project, funded by the Northern Periphery and Arctic Programme, involved partners from Finland, Sweden, Norway, Canada, Denmark/Greenland and New Zealand. It aimed to create innovative, culturally sensitive tourism products in the Arctic by sharing knowledge to better understand encounters between locals and tourists. We utilised 31 transcribed interviews conducted for the project. The interviewees were tourism entrepreneurs living in Finnish Lapland, including both Sámi and non-Sámi tourism operators from the municipalities of Rovaniemi, Kittilä, Inari, Utsjoki and Enontekiö. Some of the interviewees represented multiple stakeholders. The interview questions centred on the challenges and possibilities of utilising local cultures in tourism.

The sociomaterial practices of teaching Sámi culture were the focus of our analysis, and inductive and deductive content analysis were used to define themes and concepts in the data. The purpose of the content analysis was to organise and elicit meaning from the collected data upon which realistic conclusions could be drawn (Bengtsson, 2016). In the inductive process, conclusions drawn from the collected data wove new information into existing theory (Bengtsson, 2016). The research material was examined with an open mind by coding expressions that manifest the aspects of non-human, material, place, community and localness. A practice-based approach highlighted the relationships and networks among humans, non-humans, materials and places (Nicolini *et al.*, 2003). The four main practice categories of *doing, sensing* and *showing* everyday life and places, in addition to *storytelling* were drawn from analysis of the expressions that formed relationships between the social and material aspects and time and place.

Findings and Discussion

The importance of providing correct information to tourists about Sámi cultures was mentioned in several interviews. Correct information based on certain places and facts was seen as important. Tourism can be a valuable instrument for spreading information and supporting authenticity, local cultures and Sámi well-being. Traditional knowledge is partly tacit and goes through generations. The interviews showed that respecting traditions and local communities shaped practices, and learning is not related solely to activities, but also to social communities (Lave & Wenger, 1991). Communities of practice grow through a commitment to common practices and shared experiences. Competence is socially defined, and knowing reflects a community's socially regulated knowledge (Nicolini *et al.*, 2003) of, for example, what can be shared with tourists and by whom. A non-Sámi community member knows without saying that it is unacceptable for non-Sámi persons to represent Sámi cultures in tourism, which shows that learning evolves inside the community as well (Björn, 2020).

As one Norwegian ARCTISEN project partner put it, it is in the local DNA to know what is and is not appropriate. The Indigenous and non-Indigenous locals have learned over generations to live together in the local multi-ethnic Arctic communities, demonstrating intergenerational community internal cultural knowledge and sensitivity. Recognition is given to other community members and their culture by taking their opinions seriously in tourism business practices. What cultural sensitivity means in practice is negotiated locally (Brattland *et al.*, 2020) in relation to the local social and material aspects, place, and time (see Lüthje *et al.*, in press). For example, tourism entrepreneurs agree with reindeer herders where to run husky or snowmobile safaris in order not to disturb the reindeer or any herding procedures. However, the culturally sensitive method of acting is not clear to all tourism actors as, for instance, the touristic fake Sámi culture approves (see Kugapi *et al.*, 2020).

The ideal situation would be that more Sámi would offer tourism products based on their cultures (Björn, 2020), which would safeguard cultural heritage, strengthen the vitality of the cultures, and offer true benefits for the local community in supporting ownership of the cultures and their cultural identity. Practices, however, are not static because cultures change over time (Kuokkanen, 2010). Changes in one livelihood inevitably affect other livelihoods. As the interviews also showed, traditions can be shaped and innovated to suit modern times (Björn, 2020). The meaning of the handicraft has changed over time and changes in other livelihoods have an effect on others as well. 'There were fur shoes and such. Now we are moving on to silver jewellery, the neater work. – But of course, you have to consider that everything has changed. Reindeer herding has changed – Everything changes, and everything affects everything' (Interviewee H1). Learning in tourism should be reciprocal, and new knowledge co-created by local people (Buzinde *et al.*, 2020; Kuokkanen, 2007), tourists, and material and non-human elements provides understanding of the living nature of cultures.

In the practice-based approach, social facets include not only human, but material and symbolic elements as well. Artefacts 'participate actively in the stories, carry history, embody social relationships, distribute power, and provide points of resistance' (Nicolini *et al.*, 2003: 22–23). Material also plays an important role when teaching about Sámi cultures. Our analysis showed that tourists were taught about Sámi cultures through *doing*. This was highlighted in a Sámi handicraft workshops for tourists:

> There should be preparations, if someone wants it (to learn about Sámi culture). A teaching package including all about culture, languages, histories, nothing too big, but still the basic facts. I would like real information to be spread. It would be easier for that to happen while you do handicrafts at the same time. (Interviewee H1)

The practice of teaching culture while doing handicrafts was considered an art form in which tourists learned about the culture while creating something new and personal. Tourist interest in bringing the things learned back to their own home was mentioned by one of the interviewees. One of the tourists who had participated in a Sámi handicraft workshop wanted to start to practice similar workshops in their home country after returning there. The practice is considered social because not only do participants work together, but they work together with materials. Materials can represent a handicraft maker's cultural identity based on values and relations to nature (Björn, 2020).

According to one interviewee, handicrafts are an easy way to teach about the culture, and instruction was not only about the handicrafts, but also about language and everything related to the culture. Material elements usually raise questions in tourists, which makes it easier to teach certain things (Björn, 2020). 'If you have a product, a handicraft, or a piece of clothing from the Sámi culture with you, it always causes a lot of talking and questions' (Interviewee H2). In addition, materials carry information and memories that can be remembered long after the journey has ended (Pietikäinen & Kelly-Holmes, 2011), and they have a lot of power in Sámi dresses, which have deeply rooted cultural and symbolic meanings.

Doing in nature was also highlighted in the interviews. Making fire was mentioned as interesting for people who have not done it before. Cooking in nature can offer remarkable experiences for visitors and represent local values while teaching new skills (Björn, 2020). By being and doing in nature together with Sámi entrepreneurs, tourists also learn about the Sámi relationship with nature and how they can relate to nature in new, perhaps less human-centred, more sustainable ways (see Brattland *et al.*, 2023). For example, in the Sámi municipality of Utsjoki in northernmost Finnish Lapland, littering has become an increasing problem due to the growth in tourist numbers, so a common endeavour of local tourism actors is teaching tourists the local cultural practice of not leaving any traces of their presence in nature.

The practices of knowing and learning are not only mental but also corporeal, *multisensorial* and plural. Sensory knowledge also comprises non-human elements like technologies, organisational spaces and artefacts produced (Strati, 2007). Non-human elements concern nature and other species taking part in the processes. In the interviews, feelings were highlighted in situations where tourists experienced something new like spending a night surrounded by reindeer. Sometimes, sensations can be harder to identify such as feelings of right or wrong, which come from deeper inside (Björn, 2020). One of the interviewees mentioned that when we talk about cultural sensitivity, it is important to recognise how different things *feel*.

> When we go deeper into how things feel, even if there are foreign guides, there should be at least one Sámi or local guide who knows the customs. How to move in that nature, who to respect and where to go…/… If I see reindeer, I will turn the Safari in the other direction. These are the things that may not necessarily be taught. I'll go around or turn away. I cannot go to evict those reindeer. (Interviewee H3)

Aesthetic learning happens by combining inherent perceptions with openness to the plural world and the concept of other, which makes the knowledge co-born (Strati, 2007). This means that, while you learn from the world, you learn from yourself. The knowledge goes beyond yourself since it is born with an Other (Strati, 2007). According to one interviewee, the best learning happens when the visitor has an unexpected 'a-ha' moment. Learning situations can be experiential when visitors exceed themselves by, for example, surviving alone in the fell or forest, which evokes different kinds of sensory experiences.

Sociomaterial teaching of everyday life appeared in the interviews as *showing places* and *telling stories*. Tourism does not have to entail only made-up infrastructures; it can also show time and place through existing everyday practices. Walking helps to create productive sensory-based collective knowledge (Pink, 2008), and the practice of teaching culture by walking and sharing stories was mentioned in the interviews. Storytelling helps establish meaning and development of both individual and collective identities around non-human elements (McKenzie & Bieler, 2016), fostering a sense of belonging. Ingold (2011: 161) stated, 'In storytelling, the knowledge is integrated with the movement from place to place and from topic to topic'. According to the interviews, stories are important not only to the storytellers but also to listeners, establishing new memories. As a teaching method, storytelling helps convey history and brings different places to life, demonstrating local cultural and historical meanings. Stories can also be considered living and changing (Björn, 2020). Every place has its own story, and using humour in storytelling helps exude a more relaxed manner of teaching and learning. As one interviewee mentioned, 'people are learning types; they learn when they listen to each other'.

Conclusion

Culturally sensitive approaches to teaching tourists about local cultures include practices of doing, sensing, showing everyday life, placemaking and storytelling (see Figure 3.1). Social components, materials, time and place are entangled in the practices that demonstrate the relational quality of culturally sensitive and posthuman tourism. Reciprocal learning between tourists and locals envisions the posthuman and more sustainable future, where tourists can become active members of the learning

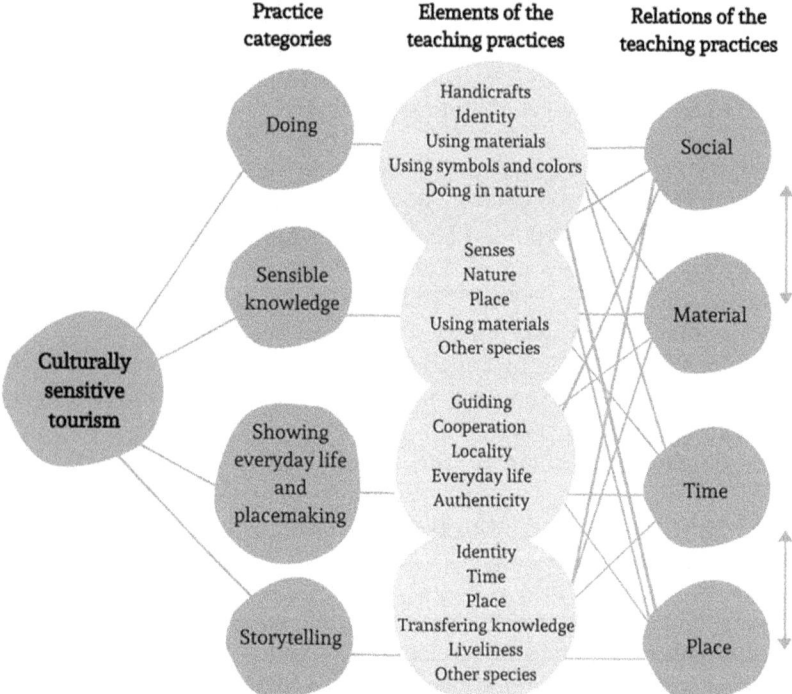

Figure 3.1 Practice categories of teaching about Sámi cultures in tourism
Source: Adapted from Björn, 2020: 48

process and be part of the community of practice. Acknowledging the role of the material and non-human in teaching practices opens new ontology and co-created knowledge. Culturally sensitive tourism products can be based on these ideologies while respecting local everyday life cultures and be planned together with the local community. New culturally sensitive tourism products promote meaningful interactions between hosts and guests, break down stereotypes and dichotomies, and enable reciprocal learning and cultural exchange enhancing mutual understanding and respect.

Sociomaterial practices of understanding different relations, cultures and places are relevant when envisioning new sustainable tourism solutions for the future. New sustainable tourism products based on everyday practices and teaching local cultures in the context of tourism require renewed attention to materiality, which may help evoke more sustainable mindsets and the rethinking of innovation (Escobar, 2018). The economy should be developed based on local needs and not grown solely for its own sake (Irwin, 2015), which highlights the role of the social innovations emerging from everyday life practices. This can be adjusted for Finnish Lapland when ideating tourism products that respect local heritage and also consider the living nature of culture being open to other networks to enhance community resilience. This requires new kinds of co-created knowledge and reciprocity among all living entities. Understanding the role of sociomaterial aspects in tourism products offers more holistic and memorable experiences for tourists to develop their cultural sensitivity by doing, sensing, listening to stories and making places.

In the case of Indigenous peoples, in particular, what is often required is recognition of the past and present destruction and exploitation of their cultures, lands, waters and resources (Hurst *et al.*, 2021; Viken *et al.*, 2021). Cultural sensitivity builds on the idea of recognising past and present painful and traumatic experiences as prerequisites for healing and reconciliation (Hurst *et al.*, 2021; Saari *et al.*, 2020), which could be supported by culturally sensitive encounters between tourists and Indigenous peoples (see Kramvig & Førde, 2020). Recognition is also prominent in ongoing academic discussion about justice in tourism (e.g. Jamal & Higham, 2021; Rastegar & Ruhanen, 2022), and we see cultural sensitivity as contributing to a more just tourism future. Posthumanist decentring of the human brings to the fore that also nature is entitled to recognition and justice (Guia & Jamal, 2023). The entanglement of nature and culture that Sámi entrepreneurs showcase sociomaterially in their products can help tourists recognise nature's entanglement with their own lives and thus contribute to a greater respect for the non-human world.

In our culturally sensitive posthumanist utopia, recognition, respect and reciprocity would be the norm to which all tourism actors would adhere. The particular characteristics of local cultures and nature would be given adequate regard and be the basis for all tourism planning and development. In tourism, there would be 'equality among socially and culturally differentiated groups, who mutually respect one another and affirm one another in their differences' (Young, 2011: 163). This utopia could be extended to other spheres of life than tourism as well. In a global, multicultural world with an ongoing environmental crisis, cultural sensitivity and a posthumanist lens would be useful everywhere. Nevertheless, it may be difficult to turn cultural insensitivity (e.g. racism, xenophobia or stereotypes) into cultural sensitivity. As Guia and Jamal (2023: 5) put it, we humans have ongoing difficulties 'to develop abilities to understand each other and create more equitable lives across differences'. Unlearning accustomed ways of thinking is not easy. There are, however, both tourists and locals who seek alternatives, which makes the emergence of heterotopias, that is, small enclaves of utopia, possible (see Matteucci *et al.*, 2022a).

Despite controversies, the Sámi municipality of Utsjoki may be one of the future heterotopias of culturally sensitive posthumanism. Nature – the non-human – is present everywhere in the small peripheral municipality and is given an important role in local cultures. The municipality has chosen cultural sensitivity as a key principle guiding its tourism planning and development. It has, among others, experimented with a local culturally sensitive collaboration model where reindeer herders and tourism entrepreneurs were brought into dialogue to plan land use together, reindeer herders also participated in revising the tourism brand of the municipality and all local residents had the possibility to participate in deciding which local sites would be promoted to tourists for visitation. Guidelines for tourists on how to take into account local nature and cultures at each site were written after discussions with the residents, local reindeer herding cooperatives, the Sámi Parliament and a multidisciplinary group of experts combining thus 'pluralistic forms of knowledge' as suggested in the new materialist tourism governance paradigm by Matteucci *et al.* (2022b: 6). These are examples of concrete steps than can be taken in other destinations as well on a path to posthumanist cultural tourism. Culturally sensitive posthumanism will be reality if we all work for it.

References

Assmann, A. (2013) Civilizing societies: Recognition and respect in a global world. *New Literary History* 44 (1), 69–91.
Bengtsson, M. (2016) How to plan and perform a qualitative study using content analysis. *NursingPlus Open* 2, 8–14.
Bennett, M. (1986) A developmental approach to training for intercultural sensitivity. *International Journal of Intercultural Relations* 10 (2), 179–196.
Birhane, A. (2021) Algorithmic injustice: A relational ethics approach. *Patterns* 2 (2), 100205.
Björn, E. (2020) Kohti posthumanistista ymmärrystä – Saamelaiskulttuurien opettaminen sosiomateriaalisena käytäntönä. [Towards sociomaterial understanding – Teaching Sámi cultures as a sociomaterial practice]. Master's thesis, University of Lapland: Faculty of Social Sciences, Tourism Research.

Brattland, C., Jæger, K., Olsen, K., Dunfjell Oskal, E.M. and Viken, A. (2020) *Cultural Sensitivity and Tourism. Report from Northern Norway.* Rovaniemi: Multidimensional Tourism Institute. See https://lauda.ulapland.fi/handle/10024/64259 (accessed April 2023).

Brattland, C., Ren, C., Bembom, E. and Bruin, R. (2023) The domestic turn in post-pandemic Indigenous Arctic tourism: Emerging stories of self and other. *Tourism, Culture & Communication* 23 (2–3), 151–162.

Buzinde, C.N., Manuel-Navarrete, M. and Swanson, T. (2020) Co-producing sustainable solutions in indigenous communities through scientific tourism. *Journal of Sustainable Tourism* 28 (9), 1255–1271.

Cerratto Pargman, T. (2023) Reconsidering learning in a socio-material world. A response to Fischer *et al.*'s contribution. *International Journal of Information and Learning Technology* 40 (1), 40–48.

Cohen, E. (2019) Posthumanism and tourism. *Tourism Review* 74 (3), 416–427.

Corradi, G., Gherardi, S. and Verzelloni, L. (2010) Through the practice lens: Where is the bandwagon of practice-based studies heading? *Management Learning* 41 (3), 265–283.

Decuypere, M. and Simons, M. (2016) Relational thinking in education: Topology, socio-material studies, and figures. *Pedagogy, Culture & Society* 24 (3), 371–386.

Escobar, A. (2018) *Designs for the Pluriverse: Radical Interdependence, Autonomy, and the Making of Worlds.* Duke University Press.

Fenwick, T., Doyle, S., Michael, M.K. and Scoles, J. (2015) Matters of learning and education: Sociomaterial approaches in ethnographic research. In S. Bollig, M. Honig, S. Neumann and C. Seele (eds) *MultiPluriTrans in Educational Ethnography: Approaching the Mutlimodality, Plurality and Translocality of Educational Realities* (pp. 141–162). Transcript Verlag/Columbia University Press.

Gherardi, S. (2011) Organizational learning: The sociology of practice. In M. Easterby-Smith and M.A. Lyles (eds) *Handbook of Organizational Learning and Knowledge Management* (pp. 43–65). Blackwell Publishing.

Guia, J. and Jamal, T. (2023) An affective and posthumanist cosmopolitan hospitality. *Annals of Tourism Research* 100, 103569.

Guttorm, H.E. (2021) Becoming Earth: Rethinking and (re-)connecting with the earth, Sámi lands, and relations. In R.-H. Andersson, B. Cothran and S. Kekki (eds) *Bridging Cultural Concepts of Nature: Indigenous People and Protected Spaces of Nature* (pp. 229–258). Helsinki University Press.

Higgins-Desbiolles, F., Carnicelli, S., Krolikowski, C., Wijesinghe, G. and Boluk, K. (2019) Degrowing tourism: Rethinking tourism. *Journal of Sustainable Tourism* 27 (12), 1926–1944.

Hirvonen, O. (2020) Johdanto – Tunnustuksen filosofia ja politiikka [Introduction – The philosophy and politics of recognition]. In O. Hirvonen (ed.) *Tunnustuksen filosofia ja politiikka. Hegelistä nykypäivään [The Philosophy and Politics of Recognition. From Hegel to Contemporary Times]* (pp. 9–22). Finnish Literature Society SKS.

Holder, A., Ruhanen, L., Walters, G. and Mkono, M. (2023) "I think … I feel …": Using projective techniques to explore socio-cultural aversions towards Indigenous tourism. *Tourism Management* 98, 104778.

Honneth, A. (1992) *Kampf um Anerkennung. Zur moralischen Grammatik sozialer Konflikte [The Struggle for Recognition. The Moral Grammar of Social Conflicts].* Suhrkamp.

Hurst, C.E., Grimwood, B.S.R., Lemelin, R.H. and Stinson, M.J. (2021) Conceptualizing cultural sensitivity in tourism: A systematic literature review. *Tourism Recreation Research* 46 (4), 500–515.

Jamal, T. and Higham, J. (2021) Justice and ethics: Towards a new platform for tourism and sustainability. *Journal of Sustainable Tourism* 29 (2–3), 143–157.

Ingold, T. (2011) *Being Alive: Essays on Movement, Knowledge and Description.* Routledge.

Irwin, T. (2015) Transition design: A proposal for a new area of design practice, study, and research. *Design and Culture* 7 (2), 229–246.

Junka-Aikio, L. (2019) Institutionalization, neo-politicization and the politics of defining Sámi research. *Acta Borealia* 36 (1), 1–22.

Junka-Aikio, L. (2022) Kulttuurisesta omimisesta uusiin identiteettikamppailuihin: Kolonialismi ja saamelaisen kulttuuripolitiikan uudet haasteet [From cultural appropriation to new identity struggels: Colonialism and the new challenges of Sámi cultural policy]. Keynote paper presented at The Starting Points of Culturally Sensitive Tourism Planning in Sámiland Seminar, 21st November 2022, University of Lapland, Rovaniemi, Finland.

Kortetmäki, T. (2020) Luontosuhteiden ja luonnon tunnustaminen [The recognition of nature relations and nature]. In O. Hirvonen (ed.) *Tunnustuksen filosofia ja politiikka. Hegelistä nykypäivään [The Philosophy and Politics of Recognition. From Hegel to Contemporary Times]* (pp. 239–254). Finnish Literature Society SKS.

Kouppanou, A. (2022) The posthumanist challenge to teaching or teaching's challenge to posthumanism: A neohumanist proposal of nearness in education. *Discourse: Studies in the Cultural Politics of Education* 43 (5), 766–784.

Kramvig, B. and Førde, A. (2020) Stories of reconciliation enacted in the everyday lives of Sámi tourism entrepreneurs. *Acta Borealia* 37 (1–2), 27–42.

Kugapi, O., Höckert, E., Lüthje, M., Mazzullo, N. and Saari, R. (2020) *Toward Culturally Sensitive Tourism: Report from Finnish Lapland*. Multidimensional Tourism Institute. See https://lauda.ulapland.fi/handle/10024/64276

Kuokkanen, R. (2007) *Reshaping the University: Responsibility, Indigenous Epistemes, and the Logic of the Gift*. UBC Press.

Kuokkanen, R. (2010) The responsibility of the Academy: A call for doing homework. *Journal of Curriculum Theorizing* 26 (3), 61–74.

Laitinen, A. (2010) On the scope of 'recognition': The role of adequate regard and mutuality. In H.-C. Schmidt am Busch and C.F. Zurn (eds) *The Philosophy of Recognition: Historical and Contemporary Perspectives* (pp. 319–342). Lexington Books.

Lave, J. and Wenger, E. (1991) *Situated Learning: Legitimate Peripheral Participation*. Cambridge University Press.

Law, J. (1992) Notes on the theory of the actor-network: Ordering, strategy, and heterogeneity. *Systems Practice* 5 (4), 379–393.

Law, J. and Urry, J. (2004) Enacting the social. *Economy and Society* 33 (3), 390–410.

Lüthje, M. (2023) Kulttuurisensitiivisyyden paikalliset käytännöt matkailuliiketoiminnassa [Local practices of cultural sensitivity in tourism business]. In S. Veijola (ed.) *Matkailunkestävä Suomi? Vastuullinen suunnittelu kulttuuri- ja luontoympäristöissä [Tourism Planning for Future: Responsible Planning in Culture-Nature Environments in Finland]* (pp. 275–312). The Finnish Literature Society SKS.

Lüthje, M., Höckert, E. and Kugapi, O. (in press) Pathways to culturally sensitive tourism policies and practices. In R. Butler and A. Carr (eds) *The Routledge Handbook of Tourism and Indigenous Peoples*. Routledge.

Magga, S.-M. (2018) Saamelainen käsityö yhtenäisyyden rakentajana. Duodjin normit ja brändit [Sámi handicrafts as the builder of unity. The norms and brands of the duodji]. Unpublished PhD thesis, University of Oulu.

Marques, L. and Oliveira, M. (2023) Promoting cultural sensitivity in higher education: An educational approach for sensitizing young travellers for local cultures. *Evolving Pedagogy*. See https://verkkolehdet.jamk.fi/ev-peda/2023/09/20/promoting-cultural-sensitivity-in-higher-education-an-educational-approach-to-sensitizing-young-travellers-for-local-cultures

Matteucci, X., Koens, K., Calvi, L. and Moretti, S. (2022a) Envisioning the futures of cultural tourism. *Futures* 142, Article 103013.

Matteucci, X., Nawijn, J. and von Zumbusch, J. (2022b) The new materialist governance paradigm for tourism destinations. *Journal of Sustainable Tourism* 30 (1), 169–184.

Maturana, H.R. and Varela, F.J. (1987) *The Tree of Knowledge: The Biological Roots of Human Understanding*. Shambhala.

McKenzie, M. and Bieler, A. (2016) *Critical Education and Sociomaterial Practice: Narration, Place, and the Social*. Peter Lang Publishing.

Nicolini, D., Gherardi, S. and Yanow, D. (2003) Introduction: Toward a practice-based view of knowing and learning in organizations. In D. Nicolini, S. Gherardi and D. Yanow (eds) *Knowing in Organizations: A Practice-Based Approach* (pp. 3–31). Routledge.

Nicolini, D. (2012) *Practice Theory, Work, & Organization: An Introduction*. Oxford University Press.

Pietikäinen, S. and Kelly-Holmes, H. (2011) The local political economy of languages in a Sámi tourism destination: Authenticity and mobility in the labelling of souvenirs. *Journal of Sociolinguistics* 15 (3), 323–346.

Pink, S. (2008) An urban tour: The sensory sociality of ethnographic place-making. *Ethnography* 9 (2), 175–196.

Rastegar, R. and Ruhanen, L. (2022) The injustices of rapid tourism growth: From recognition to restoration. *Annals of Tourism Research* 97, 103504.

Saari, R., Höckert, E., Lüthje, M., Kugapi, O. and Mazzullo, N. (2020) Cultural sensitivity in Sámi tourism: A systematic literature review in the Finnish context. *Finnish Journal of Tourism Research* 16 (1), 93–110.

Sabourin, E. (2013) Education, gift and reciprocity: A preliminary discussion. *International Journal of Lifelong Education* 32 (3), 301–317.

Sámi Parliament (2018) Principles for Responsible and Ethically Sustainable Sámi Tourism See https://www.samediggi.fi/ethical-guidelines-for-sami-tourism/?lang=en (accessed June 2023).

Strati, A. (2007) Sensible knowledge and practice-based learning. *Management Learning* 38 (1), 61–77.

Särkelä, A. (2020) Tunnustuksen patologiat. Sosiaalisuus ja kritiikki Honnethin tunnustusteoriassa [The pathologies of recognition. The social and critique in Honneth's theory of recognition]. In O. Hirvonen (ed.) *Tunnustuksen filosofia ja politiikka. Hegelistä nykypäivään [The Philosophy and Politics of Recognition. From Hegel to Contemporary Times]* (pp. 115–134). Finnish Literature Society SKS.

Turtiainen, K. (2020) Turvapaikanhakijat ja tunnustaminen [Asylum seekers and recognition]. In O. Hirvonen (ed.) *Tunnustuksen filosofia ja politiikka. Hegelistä nykypäivään [The Philosophy and Politics of Recognition. From Hegel to Contemporary Times]* (pp. 255–270). Finnish Literature Society SKS.

United Nations (1948) Universal Declaration of Human Rights. See www.un.org/en/about-us/universal-declaration-of-human-rights (accessed June 2023).

Valkonen, J. and Valkonen, S. (2014) Contesting the nature relations of Sámi culture. *Acta Borealia* 31 (1), 25–40.

van der Duim, R., Ren, C. and Jóhannesson, G.T. (2013) Ordering, materiality, and multiplicity: Enacting Actor–Network Theory in tourism. *Tourist Studies* 13 (1), 3–20.

Viken, A., Höckert, E. and Grimwood, B.S.R. (2021) Cultural sensitivity: Engaging difference in tourism. *Annals of Tourism Research* 89, Article 103223.

Walsh-Dilley, M. (2017) Theorizing reciprocity: Andean cooperation and the reproduction of community in Highland Bolivia. *The Journal of Latin American and Caribbean Anthropology* 22 (3), 514–535.

Young, I.M. (2011) *Justice and the Politics of Difference* (Paperback reissue with new foreword by D. Allen). Princeton University Press.

4 Fostering Sustainable and Resilient Rural Communities through Cultural Tourism Villages: A Case Study of the Dalmatian Hinterland

Lidija Petrić, Ante Mandić and Davorka Mikulić

Introduction

Over the past few decades, rural areas have witnessed significant economic and social transformations driven by agricultural modernisation and land management policies initiated in the 1960s (Féret *et al.*, 2020). These changes have reshaped the very definition of rurality, moving beyond the traditional focus on agricultural activities and associated lifestyles. Today, rurality is characterised by a multifaceted set of criteria encompassing administrative, demographic, locational, economic and environmental dimensions (Van Eupen *et al.*, 2012).

The evolving concept of rurality mirrors the dynamic shifts occurring in rural areas, driven by both exogenous factors like globalisation and changing social demands, as well as endogenous factors such as demographic shifts and economic diversification (Féret *et al.*, 2020). In this context, the ESPON (2020: 7) policy brief on rural regions emphasises the importance of creative mobilisation of endogenous resources as comparative territorial advantages (e.g. natural capital, local heritage, renewable energy, tourism), while the long-term vision for the EU's rural areas (European Commission, 2021) highlights actions towards stronger, connected, resilient and prosperous rural areas and communities. Notably, tourism, a key component of the tertiary sector, has played a pivotal role in diversifying the economic activities of rural regions, aligning with the sustainable tourism objectives outlined in the European transition pathway (European Commission, 2022). It has reimagined, repackaged and presented rural locales to urban markets (Rofe, 2013).

Rural areas have been repositioned as idyllic retreats, offering the allure of romance, tranquillity, freedom, nostalgia and a connection with nature – a stark contrast to the fast-paced and increasingly globalised world (Sgroi, 2022). Moreover, in the wake of global health concerns like the COVID-19 pandemic (Silva, 2021; Vaishar & Šťastná, 2020), rural areas have gained favour as safe havens. As the world becomes more complex, the demand for such rural sanctuaries continues to grow. Cultural heritage, deeply intertwined with the landscapes it inhabits, not only imparts intrinsic value but also moulds distinctive territorial identities. Moreover, when cultural heritage harmoniously converges with other contextual elements, it emerges as a catalytic force for development, as noted by Panzera (2022). Within this framework, our analysis delves into the pivotal role traditional rural villages, nurturing site-specific tangible and intangible cultural heritage can play in fostering sustainable tourism development. These villages, usually overshadowed by cultural metropolises or coastal getaways, have now evolved into primary cultural tourist attractions that hold the potential to breathe new life into the regions that cradle them. In this regard, our exploration aims to contribute to a better understanding of the rural villages' capacity for transforming into various types of cultural tourism destinations, preferably those harmonising sociocultural, environmental and economic imperatives. Their quest for equilibrium amid continuously rising multifaceted demands is of ever-mounting significance within the modern rural landscape, reflecting the evolving dynamics of rural areas, which have gradually gained popularity over urban destinations. Moreover, due to their specificities, rural communities still have an opportunity to develop regenerative tourism by adopting a bottom-up approach that is community-centred, place-based and environment-focused (Dredge, 2022).

Exploring Theoretical Frameworks for Rural Cultural Tourism Villages

In the realm of rural (cultural) tourism villages, Chiodo *et al.* (2019) have articulated a set of prerequisites for a successful transformation into sustainable tourist destinations. These villages must possess what they term 'countryside capital', comprising natural, built, economic and sociocultural assets. Furthermore, they require a commitment to sustainability from various social and economic actors and effective coordination among local stakeholders at the local and extra-local levels. Building upon this foundation, Chiodo *et al.*'s (2019) sustainable village destination model closely aligns with the concept of community-based tourism (CBT), as extensively explored in the literature (Giampiccoli & Kalis, 2012; Salazar,

2012; Giampiccoli *et al.*, 2014; Saayman & Giampiccoli, 2016; Mtapuri & Giampiccoli, 2020; Zielinski *et al.*, 2020).

CBT initiatives manifest in various forms, from community employment in businesses to full ownership and management by the local community (Dodds *et al.*, 2018). Saayman and Giampiccoli (2016) even consider independently owned rural businesses as CBT initiatives. Jugmohan and Steyn (2015) describe a model where multiple-owned structures operate under a common organisational umbrella. Yet, it is crucial to note that individual interests drive the latter CBT model, making it less suitable for pursuing rural communities' broader interests and goals (Zielinski *et al.*, 2020). Dodds *et al.* (2016) indicate that the ideal CBT model involves community ownership and management, resulting in significant benefits from small-scale local production.

In this ideal CBT model, the community leverages its diverse endogenous resources to co-create unique experiences for different types of tourists. However, this process poses a significant challenge, as communities must present an idyllic yet genuine cultural authenticity to visitors, fostering a deep connection with the local way of life (Kastenholz *et al.*, 2012). Conversely, tourists seeking to 'immerse into local culture', often described as 'purposeful tourists' (McKercher, 2002), arrive individually, eager to embrace, experience and promote community values. Notably, CBT initiatives in developing nations primarily attract foreign tourists, while rural initiatives in developed nations, due to cultural proximity, cater mainly to domestic markets, requiring less targeted marketing efforts (Zielinski *et al.*, 2020).

This delineated 'community-based tourism village model for culturally immersed tourists' aligns closely with the utopian vision of cultural tourism outlined by Matteucci *et al.* (2022a). The model prioritises Sustainable Development Goals (SDGs) and places the rights of local communities above those of tourists and tourism companies (Matteucci *et al.*, 2022a). In contrast, at the opposite end of the continuum, we find the artificial cultural tourist village – purpose-built complexes emulating specific cultural lifestyles at a particular historical period (Van Veuren, 2001). These often masquerade as 'authentic' rural villages but rely on external labour and profit-oriented entrepreneurs, families or enterprises for management (Van Veuren, 2001). Some function as museums or theme parks, targeting daily visitors, while others offer a more superficial cultural consumption experience (Simpson, 2016). This latter model, as termed by Matteucci *et al.* (2022a: 5), represents a 'dystopian model of (rural) cultural tourism'. The ethno village Herceg in Bosnia and Herzegovina (Prevolšek *et al.*, 2020) exemplifies this dystopian model, whereby culture becomes a commodity for tourist consumption rather than an intrinsic cultural asset, thus earning the label of 'profit-oriented tourism village model for sightseeing tourists'.

Between these two extremes lies a spectrum of tourism village models, each characterised by governance structures, ownership models, community involvement and the types of tourists they attract (McKercher, 2002). Sightseeing cultural tourists are often more prevalent than purposeful tourists, given that most tourists seek leisure, not cultural immersion (McKercher, 2002). However, even serendipitous tourists, initially attracted by nature, can develop deep cultural experiences when exposed to local traditions.

A wealth of literature explores rural tourism villages positioned along this continuum, especially those preserving authentic culture and incorporating them into small-scale tourism (Chiodo et al., 2019; Idziak et al., 2015; Kastenholz et al., 2012; Prevolšek et al., 2020; Qin & Leung, 2021; Sgroi, 2022; Soszyński et al., 2017; Van Veuren, 2001; Vitasurya, 2016; Yang et al., 2022). While these studies span various locations and contexts, they often share common traits, namely low population, proximity to natural and cultural assets, engagement in traditional activities, and commitment to community values and lifestyles. The UNWTO has embraced these general criteria through its Best Tourism Villages initiative, introduced in 2021 on a global scale. This initiative evaluates tourism villages against 39 criteria across 9 key areas, encompassing cultural and natural resources, sustainability, governance, infrastructure and safety. This initiative aims at promoting tourism that aligns with economic, social and environmental sustainability principles in harmony with the SDGs. As villages meet more of these criteria, they move closer to embodying the ideal UNWTO tourism village model, akin to the 'CBT village model for culturally immersed tourists'.

Recently, Sgroi (2022) documented the remarkable transformation of a traditional agricultural village into a thriving tourist hotspot celebrated for its authentic gastronomic offerings. Building upon this theme of rural transformation, Kastenholz et al. (2012) examined the case of Linhares da Beira in Central Portugal. Their work unveiled that the emotional and symbolic dimensions associated with rural experiences significantly influence tourist satisfaction, underpinned by the village's rich heritage and traditions. By way of further elaboration, Prevolšek et al. (2020) contributed additional insights through their study of six ethno-villages in Bosnia and Herzegovina. They revealed that ethno-villages with higher tourist capacities yield more significant economic benefits, but this often comes at the cost of diminished authenticity and ecological concerns. Conversely, smaller ethno-villages offer tourists a more immersive atmosphere grounded in local customs, traditions and natural resources, albeit with a more modest contribution to local community strengthening.

In Italy and Argentina, Chiodo et al. (2019) conducted multi-case-study research within various authentic rural villages involved in initiatives such as 'Borghi più belli d'Italia', 'Borghi authentic', 'Bandiere arancini' and 'Pueblos turísticos'. Their work proposed a model for analysing and monitoring the collaborative processes driving the enhancement of local assets for tourism. In 2015, Idziak et al. conducted a study on the creation of theme villages and community involvement in five thematic villages located in northern Poland. They noted that this process was distinct from the participation models proposed by Garrod (2003) or Moscardo (2008). Acknowledging the diversity of rural villages in Poland, Soszyński *et al.* (2017) provided practical spatial planning concepts tailored to different village types, from those with low summer housing intensity to mass tourism holiday resorts.

In South Africa, Van Veuren (2001) investigated artificial cultural villages, revealing that a significant portion of them was owned by white tourism operators, with only modest benefits accruing to surrounding local communities. His work also proposed strategies and activities to empower local communities and enhance their involvement in development and governance processes. Building upon this theoretical foundation, our study focuses on the Dalmatian hinterland within Split-Dalmatia County (Croatia), where we explore the current state of tourism development and the potential for diverse tourism village models. Employing a desk research methodology and drawing on our observations and involvement in development projects, we aim to provide insights into this region's unique challenges and opportunities. Due to the complexity of demographics, economics and administrative organisation, here we adopt a clustered approach, concentrating on select showcases to shape future cultural tourism development scenarios. Notably, our analysis is limited to the portion of the hinterland under Split-Dalmatia County's jurisdiction, excluding the region governed by Šibenik-Knin County.

Rural Villages' Potential for Tourism Utilisation: The Case of Dalmatian Hinterland

Common facts on Dalmatian hinterland villages

The Dalmatian hinterland, nestled within the Split-Dalmatia County, is a complex tapestry of administrative divisions. It comprises six cities and 19 municipalities, each encompassing numerous rural settlements or villages, often further divided into smaller hamlets. Our analysis focused solely on the rural settlements within these cities and municipalities, excluding those in coastal regions like Omiš and Klis. Additionally, we excluded urban-centric regions, such as the cities of Sinj, Trilj, Imotski, Vrgorac and the municipality of Dugopolje, as they exhibit more urban than rural characteristics despite their hinterland locations. This leaves us with 182 villages, home to approximately 65,198 residents, representing about 65% of the hinterland's territory and 15% of the entire Split-Dalmatia County (Croatian Bureau of Statistics, 2021).

Economically, the Dalmatian hinterland faces developmental challenges, with only four municipalities (alongside their associated villages) receiving positive rankings on the composite development index. This index classifies settlements into eight groups, with the first four indicating underdeveloped areas and the remaining four denoting developed ones (Law on Regional Development, Official Gazette, 147/14 and 123/17). From a political standpoint, the villages are closely tied to their respective cities or municipalities. However, Croatian law allows each village or settlement to establish a community committee with a council and its representative as a legal entity. This committee advocates for the village's residents within the corresponding city or municipality. It provides a platform for direct participation in decision-making processes and the co-creation of their collective future. Regrettably, the active engagement of residents, particularly the younger generation, in local projects, the development of strategic documents, or capacity-building initiatives remains a rare occurrence, primarily due to a dwindling population.

Depopulation has been an ongoing challenge in the Dalmatian hinterland since the aftermath of the Second World War. Among the 182 villages, 53 have fewer than 100 residents, 61 fall within the population range of 101 to 300 and 40 villages have populations ranging from 301 to 500 individuals. Thirteen villages boast between 500–1000 residents, 12 between 1000–2000, while only three have populations ranging from 2000–4000 (Figures 4.1 and 4.2). Consequently, 58% of the hinterland's population resides in villages with fewer than 1000 residents.

The Dalmatian hinterland is a treasure trove of historical and cultural heritage, boasting tangible and intangible facets enriching its unique tapestry. Within this intricate heritage landscape lie diverse and invaluable oral traditions, social practices, sacred rituals, festive events, indigenous ecological knowledge and the craftsmanship required for traditional crafts. Some of these cultural treasures have received well-deserved recognition

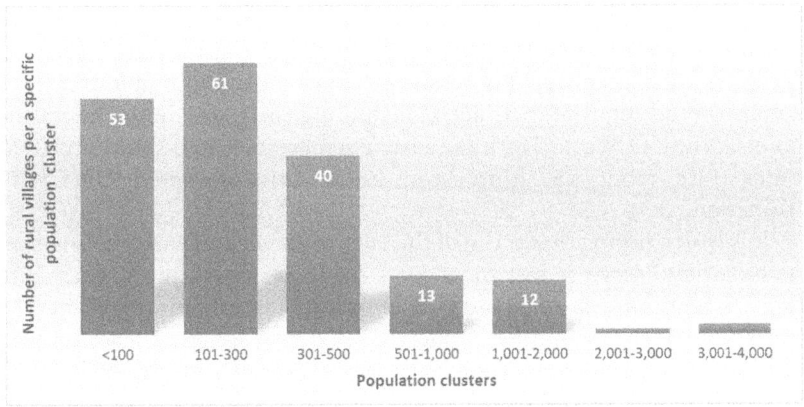

Figure 4.1 Rural villages' clusters of population
Source: Authors, based on Census 2021 by Croatian Bureau of Statistics

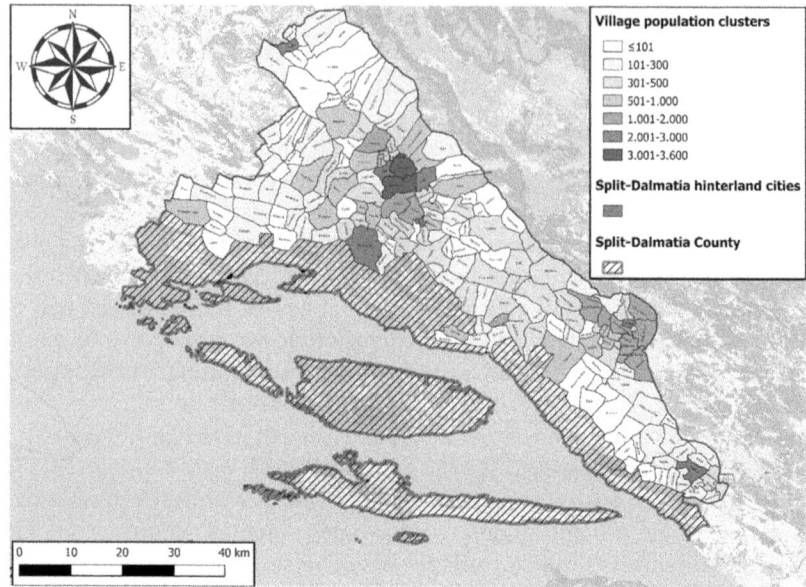

Figure 4.2 Mapping of the population-based clusters of villages in the Split-Dalmatia hinterland
Source: Authors

and protection, finding their place in national or regional registers of protected cultural assets and, in a few instances, even achieving the prestigious inclusion status in the UNESCO World Heritage List (WHL).

The hinterland's cultural wealth extends beyond celebrated archaeological sites and individual monuments. Notably, 12 rural ensembles or villages are officially recognised in the National Register of Cultural Properties, while 14 find their place in the Split-Dalmatia County Register. This preservation effort extends to the intangible cultural heritage domain, where 14 cultural assets from the hinterland area are enshrined in the National Register. Impressively, six of these intangible treasures have garnered international acclaim, finding their way onto the prestigious UNESCO World Heritage List. Furthermore, one serial archaeological site from the hinterland holds this esteemed UNESCO distinction.

It is worth noting that seven of the 26 protected rural ensembles within the hinterland serve as historical cores for larger cities like Vrgorac and Trilj, as well as a municipality, Dugopolje. The remaining protected ensembles consist of smaller hamlets nestled near larger villages or even within them, each contributing to the region's rich and diverse cultural heritage tapestry (Table 4.1).

While official protection status is noteworthy, it does not singularly determine a village's potential as an attractive tourist destination.

Table 4.1 List of protected rural ensembles in the Dalmatian hinterland

Register number	Name of a protected rural ensemble (hamlet) listed in the National Register of Cultural Properties	Location (village/ municipality)	Population of a correspondent village/municipality
Z-3852	Medvidovići	Glavina Gornja	302
Z-4088	Katići	Kozica	34
Z-3687	Veliki Godinj	Rašćane	85
Z-4086	Donji Karoglani	Zmijavci	1,654
Z-3851	Vuletića staje	Dobranje	87
Z-5360	Vrgorac	Vrgorac	5,698
Z-3587	Škopljanci	Radošić	155
Z-5596	Kukavice	Glavina Donja	1,625
Z-5419	Grubišići	Trilj	8,182
Z-6338	Zadvarje	Zadvarje	289
Z-6335	Lubine	Runović	1,706
Z-6580	Mihaljevići	Dusina	479

Register number	Name of a protected rural ensemble (hamlet) listed in the Split-Dalmatia County Register of Cultural Properties	Location (village/ municipality)	Population of a correspondent village/municipality
Ethno/E	Karadže	Cista Provo	469
E	Butige	Dicmo	2,802
E	Koprivno Ramljak	Dugopolje	3,742
E	Koprivno Dolonga	Dugopolje	3,742
E	Barići	Primorski Dolac	686
E	Šustići	Primorski Dolac	686
E	Orošnjakove Staje	Broćanac	174
E	Šeravići	Klis Konjsko	281
E	Čulići	Klis Konjsko	281
E	Meštrovići	Klis Meštrovići	281
E	Odže	Klis Odže	281
E	Kundidi	Trilj	8,182
E	Kljenak Bobanci	Vrgorac	5,698
E	Kokorići Pervani	Vrgorac	5,698

Source: Authors, based on Register of Cultural Property (https://min-kulture.gov.hr/register-of-cultural-property/16777) and Split-Dalmatia County Spatial Plan, Split-Dalmatia County Offical Gazzete, No 1/03, 8/04, 5/05 5/06 13/07, 9/13, 147/15, 154/21, 170/21

Table 4.2 Models/scenarios of tourism villages in the Dalmatian hinterland

Population clusters	<100	101–300	301–1,000	1001–4,000	>4,000	Artificial villages (no residents)
Culture/heritage characteristics						
Authentic architecture (rural ensembles	✓	✓	-	-	-	-
Fake rural ensembles	-	-	-	-	-	✓
Mixed authentic and non-authentic architecture	-	-	✓	✓	✓	-
Living intangible heritage	-	✓	✓	✓	-	-
Staged culture	✓	-	-	✓	✓	✓
Economic orientation						
Small-scale local engagement	-	✓	✓	✓	✓	-
Non-resident investors/developers	✓	-	✓	✓	✓	✓
Large tourism/recreation investments	-	-	-	✓	✓	✓
Diversified economic activities	-	✓	✓	✓	✓	-
Theme (cultural) tourism enclave	✓	-	-	-	-	✓
Governance approaches						
Profit-driven governance	✓	-	-	-	✓	✓
Community-driven governance	-	✓	✓	✓	-	-
Dominant type of visitors						
Culturally immersed/purposeful tourists	-	✓	✓	✓	-	-
Casual tourists	-	-	-	-	-	-
Incidental	-	-	-	✓	✓	-
Serendipitous	-	-	-	-	-	-
Sightseeing tourists	✓	-	-	-	-	✓

Source: Authors

Sometimes, stringent protection measures can inadvertently discourage residents from engaging in tourism-related activities due to the extensive building restrictions, effectively transforming these villages into static museums. Conversely, a lack of specific protection measures and limited spatial planning, often at the municipal level, combined with inadequate community participation, can lead to rapid and uncontrolled tourism development. This can result in sprawling, disorganised growth patterns, prioritising individual investors' interests over those of the community, as highlighted by Soszyński et al. (2017).

Regrettably, such a scenario is unfolding in the Split-Dalmatia hinterland. In less than eight years, approximately 716 houses and villas designed for tourist rentals have sprung up or have been adapted from authentic stone houses across multiple rural villages. Of these, 516 are situated in the Imotski region, 154 in the Sinj area, 47 in Trilj, 45 in Dugopolje and six in Vrlika, as stated by the local tourism boards. The appeal of these accommodations often lies in their isolation, with no residents or hosts present – a trend accentuated during COVID-19 restrictions. However, while these isolated rentals may attract visitors, they fall short of facilitating meaningful interactions with local culture. Furthermore, they fail to align with a fundamental regional policy development goal, which is to repopulate rural areas. Many property owners reside in coastal cities and rarely, if ever, meet their guests or engage in local public and community affairs.

Beyond these concerns, the low occupancy rates of such accommodations (typically ranging from 60 to 90 days per year) underscore the need to strike a balance between their numbers and the broader community's interests, thereby averting the pitfalls of spatial chaos and issues related to overtourism – challenges faced by coastal areas.

Unfortunately, this scenario is unlikely to be curbed soon, as national and regional policies actively support it. The Croatian government regularly subsidises investments in rural villa pools for rentals. At the regional level, the 'Ethno-Eco' Programme, initiated in 2004 by the Split-Dalmatia County authorities, aimed to rejuvenate abandoned and dilapidated rural villages, halting the uncontrolled real estate sales and fostering conditions for transforming these areas into appealing tourist destinations (Buble & Gamulin, 2011). This initiative also sought to establish specific governance models for each participating village, contingent upon property rights, owners' investment willingness and other distinctive factors. However, due to unresolved property rights and other impediments, the reconstruction of these villages within the programme remains incomplete, and governance models have yet to be formulated and applied. Over the years, 34 villages, including 11 in the hinterland, have entered the 'Ethno-Eco' Programme. The first decade of its implementation witnessed an expenditure of over 1 million euros, primarily allocated to infrastructure development and conservation studies. Nevertheless, this initiative failed to yield repopulation or economic diversification in the villages. Instead, only a handful of individuals, mostly non-residents, entered the rental business.

Despite the burgeoning, unregulated growth of tourist accommodations, tourism in the Dalmatian hinterland is still in its nascent stages, contributing only modestly to the County's total overnight stays – mounting to

302,799 or 1.75% (Split-Dalmatia County Tourist Board, 2023). Given the region's demographic and economic challenges, tourism presents an enticing opportunity, especially for residents who can venture into the industry with minimal investments, offering their homes or family farms. However, only 15 farms and households are currently involved in tourism (Split-Dalmatia County Board, 2023).

Tourism valorisation models in the Dalmatian hinterland villages

Drawing from the current landscape, we can discern several distinct models of tourism villages in the Dalmatian hinterland. These models diverge in their heritage classification, resident populations, resulting economic orientations, governance strategies and the types of visitors they attract (see Table 4.2 (p. 58) for a comprehensive overview).

In the case of very small authentic villages in the Dalmatian hinterland, often inhabited by ageing populations numbering less than 100 residents, the potential for 'localhood' tourism (Wonderful Copenhagen, 2020) or any other tourism endeavours is typically constrained unless external financing from non-resident investors or developers, including individuals, enterprises, or collaborations between local authorities, property owners, museums and tourist boards, comes into play. These investors often imbue the villages with thematic branding, catering predominantly to day-trippers on guided sightseeing tours, offering them interpretations of culture that, regrettably, lack authenticity. While approximately 53 villages may potentially align with this overarching model, the revitalisation and effective tourism utilisation of each necessitate a tailored approach.

Within this category of authentic rural villages, some have succumbed to complete depopulation, functioning either as *open-air museums*, exemplified by Veliki Godinj/Raščane (Figure 4.3), or adopting a *diffused eco-hotel* concept, as observed in the case of Grabovci village near Imotski

Figure 4.3 Veliki Godinj – an 'open-air museum'
Source: https://vrgorac.info/2021/04/12/veliki-godinj-2/ with permission of Vrgorac Tourist Board

(Figure 4.4). In the latter instance, the village operates under the management of homeowners, with a single individual overseeing reservations and marketing efforts. It is plausible that such models will attract growing interest, provided that depopulation trends are mitigated.

While sharing similarities with smaller counterparts, villages hosting populations ranging from 101 to 300 residents exhibit a few additional opportunities, especially when populated by younger people. Their relatively larger size allows for a greater range of possibilities, all while preserving the closely-knit communal spirit. These villages align particularly well with what we term the 'CBT village for culturally immersed tourists'.

Among the 61 villages within this population range, certain locales undoubtedly harbour untapped potential to develop along the outlined trajectory, for example, Ethno village Kokorići, situated near the city of Vrgorac (Figure 4.5). This village offers visitors an immersive experience

Figure 4.4 Grabovci (Proložac Gornji) – 'diffused eco-hotel'
Source: https://agroturizam-grabovci.com/galerija/ with permission of Agrotourism Village Grabovci

Figure 4.5 Ethno village Kokorići – CBT village for culturally immersed tourists
Source: https://vrgorac.info/2020/11/17/etno-selo-kokorici/ with permission of Vrgorac Tourist Board

in the authentic Dalmatian ambience, complemented by local gastronomy and various engaging activities involving the residents, such as strawberry, apple and grape picking, allowing them to partake in the local way of life. Development initiatives in such villages typically originate from driven individuals, as exemplified by the Kokorići village scenario, where a local hotel owner played a pivotal role in mobilising the community toward tourism-related activities.

Within the same aforementioned model, a distinct cluster comprises 53 villages with populations of up to 1000 residents. In this group, owing to the larger size than the previous category, the architectural landscape typically presents a blend of authentic rural cores and spatially scattered newly constructed structures, frequently designated for rental purposes. Despite the relatively modest population, these villages still foster robust social cohesion and a profound sense of community, which play instrumental roles in preserving and sharing their cultural heritage with visitors.

Beyond local engagement, external developers may also invest in these villages, particularly if a village has already gained recognition for specific reasons. A noteworthy illustration is the village of Zagvozd in the Imotski region, representing a 'lighter' iteration of the 'CBT model for culturally immersed tourists'. This village garnered attention due to its association with a renowned Croatian actor born there. Since 1997, he has organised an annual summer event – 'Actors in Zagvozd' – which transformed this once economically challenged village into a prominent hub for theatrical performances and ethno and popular music shows, drawing many visitors. This initiative served as a catalyst for the local community to actively engage in hosting tourists, and today, Zagvozd is celebrated for its warm hospitality and open-hearted embrace of its visitors (Figure 4.6).

As population numbers in these villages rise, the landscape and intangible heritage that once defined their authenticity have gradually given

Figure 4.6 Zagvozd – a lighter version of CBT village for culturally immersed tourists

Source: www.onlycroatia.com/travel-destinations,zagvozd with permission of Only Croatia Travel agency

way to a more urban-like character. This transformation is evident in villages such as Brnaze, Glavice and Otok in the Sinj region, where the growing population has led to a shift towards profit-oriented activities. While community involvement in governance processes may persist, the sense of close-knit belongingness tends to wane as the population expands. Over time, spurred by escalating real estate costs in coastal cities and increased demand for rural areas, individuals migrate to these villages and engage in tourism-related businesses. In some cases, these villages may even become sites for substantial tourist developments, including endeavours like golf courses in Šestanovac and Zadvarje, as indicated in the Split-Dalmatia County Spatial Plan (Official Gazette, 1/03, 8/04, 5/05, 5/06, 13/07, 9/13, 145/15 and 154/21).

While larger villages still possess tangible heritage riches, they may struggle to provide authentic, cultural, immersive experiences. Instead, they often attract sporadic tourists primarily interested in nature-based activities. To thrive, these villages would need to pivot towards developing cultural tourism on newly conceived foundations. This particular village model is aptly described as a 'culturally reinvented tourism village'. It is important to recognise that such development trends may extend beyond the 15 villages boasting populations exceeding 1000 residents. Other villages could follow a similar trajectory if there are no specific government policies to preserve their authenticity.

Lastly, there is the model of artificial villages devoid of permanent residents. These are profit-oriented ventures, functioning either as theme museums or hotel resorts, offering staged authenticity for sightseeing tourists. Despite the lack of permanent inhabitants, they can serve as effective vehicles for showcasing the region's heritage. A notable illustration is the village known as Stella Croatica within the Klis municipality (Figure 4.7). It forms part of a larger complex owned by a family primarily

Figure 4.7 Stella Croatica – non-authentic profit-oriented model of a tourist village
Source: www.dalmatia.hr/attractions/the-stella-croatica-ethno-agro-park with permission of the Stella Croatica – Stella Mediterranea d.o.o.

producing indigenous olive oil, marmalades, cookies, honey and herbal cosmetics. The village, accompanied by a botanical park featuring over 500 authentic herbs, is meticulously designed to interpret the traditions and culture of Dalmatian hinterland villages. While the project has demonstrated success, it could offer an even richer experience for tourists and the local community if the traditions were presented by those who originally cultivated them in their original settings.

Policy Implications and Future Outlook

Aligning policy implications with UNWTO's Best Tourism Villages principles

A paramount consideration in shaping tourism policies for the Dalmatian hinterland is the delicate equilibrium between heritage preservation and sustainable development. The region's attraction lies in its genuine cultural and natural resources, making it crucial to adopt policies that promote responsible tourism. These practices should be rooted in celebrating the region's unique character while mitigating the risks of overdevelopment that might compromise its authenticity, thus aligning with the UNWTO's call for conservation and promotion of cultural and natural resources, particularly in rural regions. We propose tailored strategies for each category, depending on contextual factors (such as population density and resulting issues associated with the use of space, sense of identity, etc.), mindful of the need to balance heritage preservation and sustainable development, as underscored by the region's unique characteristics and the criteria put forth by UNWTO (2021) for Best Tourism Villages.

Preserving heritage in very small villages

For very small villages, typically inhabited by ageing populations with fewer than 100 residents, policies should focus on encouraging partnerships between local authorities, property owners, museums and tourist boards. These partnerships could attract external financing, allowing for thematic branding and catering primarily to day trippers on guided tours. However, it is crucial to emphasise the importance of offering authentic cultural interpretations to enhance the visitor experience. In cases where depopulation has occurred, policies should promote the development of open-air museums or diffused eco-hotels. Still, these initiatives would need to be accompanied by measures to address depopulation trends effectively.

Engaging communities and balancing sustainability in small and mid-sized villages

Policies should prioritise community involvement in villages hosting populations ranging from 101 to 300 residents, especially those boasting a younger demographic and a closely-knit communal spirit. Initiatives aimed at capacity building, educational programs and incentives can stimulate residents' active participation in shaping the tourism narrative. Encouraging motivated individuals within these communities to take the lead in initiating tourism-related activities and in engaging residents in community activities can be particularly effective. These policies align with UNWTO's emphasis on fostering community engagement and promoting human resources development. Mid-sized villages with populations ranging between 301–1000 residents enjoy a mix of authentic rural cores and scattered newly constructed structures. The sense of community in these villages tends to be fading away gradually. Therefore, aside from promoting community engagement, policies should prioritise enhancing authenticity and sustainability. Regulations must be in place to strike a harmonious balance between accommodating tourists and preserving the region's authentic character. These policies align with UNWTO's (2021) focus on responsible tourism practices and tailored approaches to preserving cultural resources.

Reinventing cultural tourism by diversifying offerings in large villages

For large villages with over 1000 residents, which may struggle to provide authentic, cultural, immersive experiences, policies should encourage the diversification of tourism offerings beyond conventional accommodation options. Culture, tradition and ecotourism experiences should be nurtured and effectively marketed to cater to a wider range of visitor interests while preserving the region's authentic character. Considering the existing structures and environment, clear regulations and zoning laws concerning tourist accommodations and large-scale recreational facilities (such as golf courses) must be in place. These policies align with UNWTO's (2021) emphasis on responsible tourism practices and promoting innovative products and experiences.

Enhancing visitor experiences in artificial villages

In the case of artificial villages devoid of permanent residents, policies should support their effective use as theme museums or diffused eco-hotel resorts while ensuring that they effectively showcase the region's heritage. These villages offer opportunities for investments in interactive technologies (AI and VR) to elevate the visitor experience. Policymakers should also explore opportunities for local involvement in interpreting traditions and culture within these artificial villages, potentially enriching the visitor experience by involving the community.

Future outlook

The Dalmatian hinterland holds immense potential for sustainable tourism development, aligning with the UNWTO's (2021) vision for the Best Tourism Villages. By preserving and celebrating its unique cultural and natural assets, the region would attract responsible tourists seeking authentic and immersive experiences that positively impact the environment and local communities. Integrating technology into tourism, such as VR experiences and digital storytelling, may elevate the visitor experience while preserving heritage sites. Policies should support the widespread adoption of such technologies to engage and educate tourists effectively, enriching their understanding of the region's history and culture.

Initiatives promoting cultural exchange between urban and rural areas can foster cohesion and bridge societal divides. These programs provide a platform for residents to participate in tourism development actively, ultimately enhancing the visitor experience by showcasing the region's unique traditions and heritage. Collaboration among municipalities, counties and national agencies is pivotal for holistic tourism development. By pooling resources, coordinating marketing efforts and aligning policies within and across sectors and institutions, the region would optimise its potential and overcome challenges more effectively, fostering a sense of unity among stakeholders.

Investment in education, training programs and initiatives related to tourism and heritage preservation is a fundamental long-term strategy. Equipping residents with the necessary skills and knowledge would empower them to participate in tourism development actively, ensuring sustainability. However, arming residents with knowledge and skills is not enough to achieve sustainable cultural tourism development. There is a need to adopt an alternative governance paradigm for tourism destinations. In this regard, Matteucci *et al*. (2022b) have called for the adoption of a governance framework inspired by new materialist philosophy. Such a governance framework entails a post-anthropocentric worldview, participatory modes of decision-making, resilient forms of cultural tourism activities (e.g. creative tourism), and social eudaimonia (or community well-being) as societal value.

Regular monitoring of the impact of tourism on the region, including social, economic and environmental factors, is indispensable. Policies should be adaptive, responsive to emerging trends and challenges, and geared toward long-term sustainability. Learning from the COVID-19 pandemic, tourism policies should incorporate contingency plans for crises. Diversification of tourism offerings can mitigate the impact of sudden shocks, ensuring resilience in the face of adversity. The hinterland's rich cultural heritage offers opportunities for cultural exchange programs and partnerships with international organisations. Effective marketing and promotion of these initiatives would attract a global audience interested in heritage preservation and authentic experiences. However, the region must be mindful of the risk of overtourism and its strain on cultural resources. It is therefore crucial to ensure that tourism development does not compromise the very cultural and natural assets that draw visitors in the first place. On the flip side, there are significant opportunities, some of which have been identified above, including leveraging technology for sustainable tourism management, tapping into international collaborations for heritage preservation, and mobilising local communities for authentic and immersive tourist experiences. Effective policies would capitalise on these opportunities to not only address challenges but also to elevate the region's tourism appeal.

Conclusion, Limitations and Future Research

This study has comprehensively explored the Dalmatian hinterland's potential for tourism development. We have examined the intricate interplay between heritage preservation and sustainable tourism. Subsequently, we have scrutinised the unique characteristics of the region's villages. These insights have led to critical policy implications and a promising future outlook for this, culturally and ecologically rich region. Although this chapter offers valuable insights, our study has certain limitations. While the empirical analysis is comprehensive, it may not fully capture the evolving dynamics within the region as trends are shaped by multiple sectors outside tourism. Furthermore, while our findings may only be applicable to the Dalmatian's hinterland, they provide the foundation for a valuable benchmark for future studies.

As far as future research is concerned, a comparative analysis of rural tourism development across regions would provide insights into policy effectiveness and best practices. Also, longitudinal studies tracking the evolution of tourism in the Dalmatian hinterland would provide a deeper understanding of the region's resilience and adaptability. Research on tourist behaviour and preferences in the hinterland would guide the development of tailored experiences. Because assessing the environmental impact of tourism is crucial for ensuring sustainability, ongoing evaluation of tourism policies and their impact on the region's development and heritage preservation would be worth pursuing. Furthermore, in light of the COVID-19 pandemic, exploring crisis management strategies for rural tourism destinations would facilitate preparedness. Finally, with respect to tourism governance, research on the dynamics involved in mindsets change would be a first step towards regenerative tourism future, not only in this region but in Croatia in general. We hope that this study will serve as a stepping stone towards well-crafted policies, which will aim at realising the full potential of our beautiful region.

References

Buble, S. and Gamulin, A. (2011) Prostorno planiranje u ruralnim cjelinama. *Klesarstvo i Graditeljstvo* 21 (1–2), 90–103.

Croatian Bureau of Statistics (2021) Census 2021.

Chiodo, E., Adriani H.L., Navarro, F.P. and Salvatore, R. (2019) Collaborative processes and collective impact in tourist rural villages—Insights from a comparative analysis between Argentinian and Italian cases. *Sustainability* 11 (2), 432.

Dodds, R., Ali, A. and Galaski, K. (2018) Mobilising knowledge: Determining key elements for success and pitfalls in developing community-based tourism. *Current Issues in Tourism* 21 (13), 1547–1568.

Dredge, D. (2022) Regenerative tourism: Transforming mindsets, systems and practices. *Journal of Tourism Futures* 8 (3), 269–281.

ESPON (2020) Policy Brief, Shrinking Rural Regions in Europe. Towards Smart and Innovative Approaches to Regional Development Challenges in Depopulating Rural Regions. ESPON EGTC.

European Commission (2022) *Transition Pathway for Tourism*, Directorate-General for Internal Market, Industry, Entrepreneurship and SMEs, Brussels. See https://single-market-economy.ec.europa.eu/news/transition-pathway-tourism-published-today-2022-02-04_en (accessed June 2023).

European Commission (2021) *A Long-Term Vision for the EU's Rural Areas – Towards Stronger, Connected, Resilient and Prosperous Rural Areas by 2040*. Communication from the Commission to the European Parliament, The Council, The European Economic and Social Committee and the Committee of the Regions, COM(2021)345 final.

Féret, S., Berchoux, T., Requier, M. and Abdelhakim, T. (2020) D3.2 Framework providing definitions, review and operational typology of rural areas in Europe. CIHEAM-IAMM. H2020 Call ID & Topic RUR-01-2018-2019–D / Rural society-science-policy hub, SHERPA: Sustainable Hub to Engage into Rural Policies with Actors, Project ID 862448

Giampiccoli, A. and Kalis, J.H. (2012) Community-based tourism and local culture: The case of the AmaMpondo. *PASOS Revista de turismo y patrimonio cultural* 10 (1), 173–188.

Giampiccoli, A., Jugmohan, S. and Mtapuri, O. (2014) International cooperation, community-based tourism and capacity building: Results from AmaMpondoland village in South Africa. *Mediterranean Journal of Social Sciences* 5 (23), 657–667.

Idziak, W., Majewski, J. and Zmyślony, P. (2015) Community participation in sustainable rural tourism experience creation: a long-term appraisal and lessons from a thematic villages project in Poland. *Journal of Sustainable Tourism* 23 (8–9), 1341–1362.

Jugmohan, S. and Steyn, J.N. (2015) A pre-condition evaluation and management model for community-based tourism. *African Journal for Physical Health Education Recreation and Dance* 21 (3.2), 1065–1081.

Kastenholz, E., Carneiro, M.J., Peixeira, M.C. and Lima, J. (2012) Understanding and managing the rural tourism experience – The case of a historical village in Portugal. *Tourism Management Perspectives* 4, 207–214.

Law on Regional Development, *Official Gazette*, 147/14 and 123/17.

Matteucci, X., Koens, K., Calvi, L. and Moretti, S. (2022a) Envisioning the futures of cultural tourism. *Futures* 142, Article 103013.

Matteucci, X., Nawijn, J. and von Zumbusch, J. (2022b) A new materialist governance paradigm for tourism destinations. *Journal of Sustainable Tourism* 30 (1), 169–184.

McKercher, B. (2002) Towards a classification of cultural tourists. *International Journal of Tourism Research* 4 (1), 29–38.

Mtapuri, O. and Giampiccoli, A. (2020) Beyond rural contexts: Community-based tourism in urban areas. *Advances in Hospitality and Tourism Research* 9100, 419–439.

Register of Cultural Properties. See https://min-kulture.gov.hr/register-of-cultural-property/16777 (accessed June 2023).

Panzera, E. (2022) *Cultural Heritage and Territorial Identity: Synergies and Development Impacts on European Regions*. Advances in Spatial Science. The Regional Science Series, Springer Nature Switzerland AG.

Prevolšek, B., Maksimović, A., Puška, A., Pažek, K., Žibert, M. and Rozman, Č. (2020) Sustainable development of ethno-villages in Bosnia and Herzegovina: A multi criteria assessment. *Sustainability* 12 (4), Article 1399.

Qin, R.J. and Leung, H.H. (2021) Becoming a traditional village: Heritage protection and livelihood transformation of a Chinese village. *Sustainability* 13 (4), Article 2331.

Rofe, M.W. (2013) Considering the limits of rural place-making opportunities: Rural dystopias and dark tourism. *Landscape Research* 38 (2), 262–272.

Saayman, M. and Giampiccoli, A. (2016) Community-based and pro-poor tourism: Initial assessment of their relation to community development. *European Journal of Tourism Resources* 12, 145–190.

Salazar, N.B. (2012) Community-based cultural tourism: Issues, threats and opportunities. *Journal of Sustainable Tourism* 20 (1), 9–22.

Sgroi, F. (2022) Evaluating the sustainability of complex rural ecosystems during the transition from agricultural villages to tourist destinations and modern agri-food systems. *Journal of Agriculture and Food Research* 9, Article 100330.

Silva, L. (2021) The impact of the COVID-19 pandemic on rural tourism: A case study from Portugal. *Anatolia* 33 (1), 157–159.

Simpson, T. (2016) Tourist utopias: Biopolitics and the genealogy of the post-world tourist city. *Current Issues in Tourism* 19 (1), 27–59.

Soszyński, D., Sowińska-Świerkosz, B., Stokowski, P.A. and Tucki, A. (2017) Spatial arrangements of tourist villages: Implications for the integration of residents and tourists. *Tourism Geographies* 20 (5), 770–79.

Split-Dalmatia County Spatial Plan. Split-Dalmatia County Official Gazzette, 1/03, 8/04, 5/05 5/06 13/07, 9/13, 147/15, 154/21, 170/21.

Split-Dalmatia County Tourist Board (2023) Statistical analysis of tourism traffic in 2022. See https://www.dalmatia.hr/wp-content/uploads/2023/04/Analiza2022.pdf (accessed June 2023).

UNWTO (2021) *Best Tourism Villages by UNWTO, Areas of Evaluation*, See www.unwto.org/tourism-villages/en/ (accessed May 2023).

Vaishar, A. and Šťastná, M. (2020) Impact of the COVID-19 pandemic on rural tourism in Czechia: Preliminary considerations. *Current Issues in Tourism* 25 (2), 187–191.

Van Eupen, M., Metzger, M.J., Pérez-Soba, M., Verburg, P.H., Van Doorn, A. and Bunce R.G.H. (2012) A rural typology for strategic European policies. *Land Use Policy* 29 (3), 473–482.

Van Veuren, E.J. (2001) Transforming cultural villages in the spatial development initiatives of South Africa. *South African Geographical Journal* 83 (2), 137–148.

Vitasurya, V.R. (2016) Local wisdom for sustainable development of rural tourism, case on Kalibiru and Lopati Village, Province of Daerah Istimewa Yogyakarta. *Procedia – Social and Behavioral Sciences* 216, 97–108.

Wonderful Copenhagen (2020) The End of Tourism as we Know it – Towards a New Beginning of Localhood. Strategy 2020. See http://localhood.wonderfulcopenhagen.dk/wonderful-copenhagen-strategy-2020.pdf (accessed April 2023).

Yang, J., Ma, H. and Weng, L. (2022) Transformation of rural space under the impact of tourism: The case of Xiamen, China. *Land* 11, Article 928.

Zielinski, S., Jeong, Y., Kim, S. and Milanés, C.B. (2020) Why community-based tourism and rural tourism in developing and developed nations are treated differently? A review. *Sustainability* 12 (15), 5–18.

5 Uncovering the Multifaceted Heritage Values of Longhushan World Natural Heritage Site through Tourists' Lens: An Analysis of Online Travelogues

Rouran Zhang, Weili Zhan, Ying Lyu and Da Kuang

Introduction

The nomination of World Heritage Sites is a complex, two-fold process. Firstly, it involves the analytical work of experts who identify and substantiate the 'Outstanding Universal Value' (OUV) of a site while ensuring its authenticity and integrity, following the Operational Guidelines for the Implementation of the World Heritage Convention (henceforth, 'the Guidelines'). Secondly, the process requires political diplomacy, necessitating constant communication and consultation among UNESCO, the governments of the State Parties, and the World Heritage advisory bodies such as the International Council on Monuments and Sites (ICOMOS) and the International Union for Conservation of Nature (IUCN) (Zhang & Brown, 2022).

In August 2010, during the 34th session of the World Heritage Committee, 'China Danxia' (中国丹霞) was inscribed as the second successful natural World Heritage Site from China following a collaborative nomination. An important element of this joint endeavour was Longhushan in Jiangxi Province. Reflecting on the journey from its initial controversy to eventual triumph, the successful nomination of 'China Danxia' was the result of a dedicated expert team who worked diligently to articulate its OUV and secure global acceptance (Tang & Wan, 2010).

Furthermore, at a national level, effective diplomatic negotiations garnered the support of a majority of World Heritage Committee member states, leading to the overturning of IUCN's deferral recommendation and the site's successful inscription on the World Heritage List (Li & Chen, 2010).

Despite these achievements, the 'China Danxia' nomination process predominantly focused on the site's natural value, unintentionally downplaying the rich cultural significance of Longhushan as a revered Taoist site. This neglect reveals a common issue associated with the nomination process of heritage sites, namely the lack of multi-stakeholder inclusiveness in heritage-making. Heritage sites like Longhushan, due to their fame and unique natural and cultural values, have become prominent tourist attractions. Tourists, as the main participants in tourism activities, engage in a comprehensive interpretation, reconstruction and retransmission of the meaning of the tourist destination throughout the entire tourism process. This chapter interprets the diverse values of Longhushan from the perspective of tourists through the analysis of online discourse. It aims to compare authoritative discourse and the understanding of Mount Longhu's heritage value by Chinese scholars with the viewpoints of tourists. The goal is to emphasise the cultural importance of Longhushan, highlighting its indispensable role in the integration of culture and nature. This is intended to correct the official oversight of Longhushan's unique cultural value, highlight the importance of incorporating public discourse into the future development of cultural tourism, and serve as a reference for constructing a more resilient and inclusive utopian vision of cultural tourism.

Cultural Tourism Futures

Heritage is experienced, valued and enjoyed by a variety of stakeholders such as cultural groups who live near a heritage asset as well as domestic and international visitors. In fact, in the last decades, du Cros and McKercher (2015) note, tourism actors have become key stakeholders in the practice of cultural heritage management. Heritage-making, to which the nomination of World Heritage Sites is related, is a fluid process, understood as 'performance' (Smith, 2015), constructed by those who are attached to an asset culturally (e.g. locals, tourists). In heritage-making, however, frictions among multilayered stakeholders are inevitable, which present challenges for the inclusion of an asset on the World Heritage List.

Matteucci *et al.* (2022) proposed three visions of the future development of cultural tourism: utopian, dystopian and heterotopian vision. Utopian cultural tourism envisions an ideal future where sustainability, equity and community well-being are at the forefront. This concept aims to create harmonious tourism experiences that benefit both local communities and visitors while preserving cultural heritage and minimising environmental impact. In a utopian vision, not only those who belong to an international elite, but also a plethora of 'weaker' voices, all infused with ethical and democratic concerns, would enter into dialogue, and exchange information for the common aim to evaluate and nominate World Heritage Sites. In contrast, a dystopian vision represents a bleak outlook where negative trends dominate, resulting in the degradation of cultural and environmental resources, deepening social inequalities and the commodification of culture. This scenario portrays a tourism landscape marked by conflicts, exploitation and loss. In that respect, Harrison (2021: 128) reminds us that 'heritage is linked to *power*: the power to impose a view of the world, especially of the past, on others. Perceptions of the past are closely linked to present hierarchies, and the voices of those at the top are often the most likely to prevail'. On the other hand, heterotopian cultural tourism describes an alternative approach where small-scale, resistant spaces emerge within the prevailing system. These 'pockets of resistance' focus on ethical practices, social justice, and cultural diversity, offering a counter-narrative to mainstream tourism and creating spaces for alternative, more equitable tourism practices. By inspiring many communities and policymakers, Matteucci *et al.* suggest that these pockets of resistance will multiply, thus their models of governance may become dominant, utopian practices.

World Heritage Nomination Process and Value Evaluation

The profound social transformations and widespread destruction wrought by the two World Wars prompted Western European countries to prioritise the safeguarding of tangible heritage assets, such as architectural structures, archaeological sites and natural landscapes. Propelled by international organisations like UNESCO, various post-war conventions and charters were instituted, fostering a global movement for the protection of the 'common heritage' of mankind (Yu & Zhang, 2020).

In 1972, the Convention Concerning the Protection of the World Cultural and Natural Heritage (hereafter 'the Convention') formalised the Western approach to heritage protection, predicated on the concept of materiality. This framework segregated heritage into two distinct domains: cultural and natural. The 10 criteria of OUV outlined in the Operational Guidelines direct the specific practices of World Heritage nomination, management, and related matters. As per the 2021 revised version of the Operational Guidelines, the nomination process for World Heritage is represented in Figure 5.1.

From the outlined process, we can see that the World Heritage nomination procedure is orchestrated by representatives and experts from the contracting states, the World Heritage Committee, ICOMOS and IUCN. Within this authoritative heritage framework, the nomination of World

Uncovering the Multifaceted Heritage Values of Longhushan World Natural Heritage Site   73

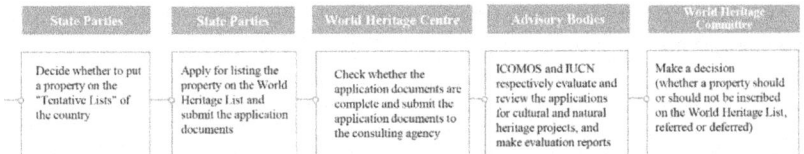

Figure 5.1 World Heritage Nomination Process (UNESCO, 2021a)
Source: Self-drawn based on the revised version of the Operational Guidelines in 2021

Heritage Sites has evolved into an international platform allowing countries to exhibit their soft power. National and local governments highly value the political and economic benefits that accompany the inclusion of heritage sites on the World Heritage List. As a result, during the nomination process, inaccuracies may occur in assessing the OUV of heritage sites due to knowledge limitations of the ICOMOS and IUCN representatives and experts. Moreover, governments may utilise diplomatic means to influence the World Heritage Committee's evaluation results, given the associated political and economic gains (Liuzza & Meskell, 2021; Meskell *et al.*, 2015; Winter, 2015). These situations can potentially compromise the authenticity and integrity of heritage value interpretations (Zhang & Brown, 2023). This chapter primarily addresses the former, exploring the limitations of IUCN's value assessment in the nomination of 'China Danxia' as a World Heritage site and investigating heritage site values from a public perspective.

At the dawn of the 21st century, leading heritage scholar Laurajane Smith introduced the concept of 'Authorised Heritage Discourse' (AHD). AHD sought to critique heritage conservation practices that were Eurocentric, focusing on expert discourse and materiality of heritage, contending that such practices obstructed the inclusion of other stakeholder groups with heritage interests, including women and ethnic minorities (Smith, 2006). Smith proposed that heritage is a sociocultural process with its value created through the engagement of various stakeholders, thus underscoring its diversity (Smith, 2020). In a utopian future, heritage conservation would therefore involve a multiplicity of civil society groups such as indigenous peoples and other minorities.

In practice, the UNESCO World Heritage program has gradually integrated new concepts. For instance, in 1992, 'cultural landscapes' were incorporated as a new category in the operational guidelines, acknowledging the value of culture-nature symbiosis (Zhang, 2020). The 1994 Nara Document on Authenticity advocated for attention to the authenticity and diversity of heritage values (ICOMOS, 1994). Moreover, in 2021, the Fuzhou Declaration, announced at the 44th World Heritage Committee session, further emphasised the necessity of the organic integration of cultural and natural values. It also advocated for broader participation from academia, civil society and local communities in heritage conservation, management and evaluation (UNESCO, 2021b). Therefore, it is clear that both academia and heritage practice underline the importance of collaborative interpretation of heritage values by a variety of stakeholders.

Research on the heritage value of Longhushan

Longhushan, situated in Yingtan City, Jiangxi Province, China, is recognised as the birthplace of Taoism in the country. It showcases a distinctive Danxia landform (丹霞地貌) characterised by scattered peaks and expansive valleys with solitary peaks. In 2008, it was inscribed on the World Heritage List as part of the 'China Danxia' site. 'China Danxia' denotes a natural landscape of red sandstone formations that evolved in warm, humid monsoon climates. UNESCO's assessment found that 'China Danxia' satisfies criteria vii and viii (Table 5.1) of the Operational Guidelines.

At present, Chinese scholars typically explore the value of Longhushan from three perspectives: its natural attributes, tourism development and Taoist culture. For instance, they analyse the geological characteristics of Longhushan, such as its geological structures, paleontological formations, sedimentary rock attributes and Danxia landform features, to understand its natural and scientific value (Guo *et al.*, 2013; Huang, 2010; Jiang *et al.*, 2006; Liao, 2019; Zhu *et al.*, 2012). Additionally, they delve into the process of tourism development at Longhushan and examine the relationship

Table 5.1 Analysis on the authoritative Evaluation of the Longhushan Heritage (UNESCO, 2010)

Heritage value	Specific standard		Specific elements
Natural value	Aesthetic Value (criteria vii)	An extraordinary natural phenomenon or an area of special natural beauty and aesthetic importance.	Red conglomerate and sandstone make up this unique natural beauty, forming spectacular peaks, pillars, cliffs and majestic canyons
	Scientific value (criteria viii)	Outstanding examples that represent major stages of the earth's history, including life records and important geological processes in the development of topography	Major geological events such as early Cretaceous volcanic activity, late Cretaceous gypsum salt deposition, aeolian sand accumulation and dinosaur catastrophe in the Xinjiang basin where Longhushan is located record the important geological evolution of the Cretaceous in this area
		Important geomorphological or topographic features	The outstanding characteristics of erosion residual peak clusters, peak forests, isolated peaks and residual hills indicate that it belongs to the representative of Danxia type of wide valley type of evacuated peak forest from late adult to early old age

Table 5.2 Analysis of the value of Longhushan heritage evaluated by Chinese scholars

Heritage value		Specific elements
Natural value	Aesthetic value	The model of natural beauty of Bishui Danshan landscape, the natural copy of traditional Chinese landscape painting; Danxia pictographic stone peak and its combination are models of rare and micro geomorphological landscapes in the world.
	Scientific value	There are many tortoise crack landscapes developed in Danxia landform in Longhushan area of Jiangxi Province. The Luxi River coastal zone of Longhu Mountain is a typical representative of the aged Danxia peak forest, and it is also a model of river erosion evacuation type Danxia wide valley peak forest. Longhushan Mazuyan and Golden Gun Peak are typical examples of Danxia solitary peak-hilly landforms in the old age; Popular science education in Longhushan World Geopark is of great help for tourists to familiarise themselves with geoscience knowledge.
	Biological and ecological diversity value	Longhu Mountain is a typical representative of low-altitude natural vegetation of Danxia landform in the middle subtropics of the world, is an important breeding ground for birds in Southeast Asia, and is rich in wildlife resources.
Cultural value		Cliff Tomb Culture: The cliff tombs of Longhu Mountain were distributed in groups more than 2,600 years ago. The cliff tombs are steep and unattainable, which has become an eternal mystery. Taoist culture: Longhushan is the birthplace and inheritance place of Taoism. It is a holy place for Taoist believers at home and abroad to seek immortals, visit ancestors and make pilgrimages.

between residents and tourists in the region (Fu & Yu, 2019; Guo et al., 2012; Xiaoqian et al., 2018; Zhang & Guan, 2022). Through analysing Longhushan's cultural value, particularly its significance in Taoist culture, they underscore the role of cultural value in boosting tourism (Chen, 2009; Yao, 2020). Table 5.2 encapsulates the research on Longhushan's heritage value conducted by domestic scholars.

Currently, the interpretation of Longhushan's heritage value mainly centres on two dimensions: official application documents and studies undertaken by domestic scholars, with limited analysis from a public viewpoint. This chapter aligns with the recent international emphasis as advocated in the Fuzhou Declaration, which calls for a bottom-up approach to discover the diverse values of heritage construed by different stakeholders. It aims to uncover the heritage value of Longhushan from a tourist perspective through network text analysis.

Research Method

This study utilises the ROST-CM6[1] to process and analyse tourist comments on travel websites and visualises the findings using Gephi[2]. The objective is to capture and articulate tourists' perceived value of

Table 5.3 Review data obtained by travel websites

Website	Number of comments/posts
https://www.ctrip.com/	2768
https://www.mafengwo.cn/	21
Total	2789

Longhushan in Jiangxi and contrast it with the OUV ascribed to Longhushan as a World Heritage site, as well as with the heritage value investigated by domestic scholars. For this purpose, we selected two renowned travel websites, Ctrip (携程旅游网) and Mafengwo Travel Website (马蜂窝旅游网), to gather tourist comments on Longhushan using the Octoparse software. The comment information is collected from January 1, 2015 through May 5, 2022.

During the preliminary cleaning of raw textual data, incomplete comment records are eliminated, incorrect expressions are rectified, and comments that include no text, only images, mere scenic introductions, or repetitive invalid comments are discarded. Eventually, a total of 2789 valid evaluations (Table 5.3, coded LHS0001-LHS2789) are organised, offering nearly 180,000 words of text for analysis.

Research Findings: Semantic Analysis of Web Text

Prior to employing ROST-CM6 for word frequency analysis, we pre-processed the text content by (1) adding related professional terms such as 'Elephant Trunk Mountain (象鼻山), Danxia Landform (丹霞地貌), Zhang Daoling (张道陵[3]), Xiao Guilin (小桂林[4]), Upper Pure Ancient Town (上清古镇[5])' into the custom dictionary to ensure accurate word separation and the effectiveness of subsequent analysis, and (2) according to the results of word frequency statistics, correcting misexpressed words and merging semantically similar words.

The words frequently appearing in the text represent tourists' focus areas (Zeng *et al.*, 2022). This study extracts the top 100 high-frequency words from the text of tourists' comments and endeavours to construct a value classification of Longhushan from the tourists' viewpoint. Words like 'Longhushan, Jiangxi (江西), Yingtan (鹰潭), China' merely state the subject and location information, while 'time, then, afternoon, old man, kilometer, and also' are unrelated to the research content and are, therefore, excluded from the count. The remaining 90 high-frequency words are then classified according to their semantic meanings (Table 5.4), with 37.09% relating to tourism activities and 62.91% directly or indirectly reflecting the perceived heritage value.

The 56 high-frequency words that denote the heritage value are further detailed to explore the heritage value from the tourists' standpoint,

Table 5.4 Semantic classification of high-frequency words

Semantic classification	The number of high-frequency words (number)	Total occurrence frequency (number)	Proportion (%)
Related to heritage value	56	9071	62.91
Related to tourism activities	34	5348	37.09

Table 5.5 Heritage value classification based on high-frequency words

Heritage value	High-frequency words
Cultural value (3162/34.86%)	hanging coffin (悬棺), Upper Pure Ancient Town , Upper Pure Palace (上清宫[6]), ancestral court(祖庭), heritage, history, cliff tomb (崖墓), World Natural Heritage, Eastern Han Dynasty (东汉), Taoism, culture, Celestial Master's Mansion (天师府), Tianshi (天师[7]), no mosquito village (无蚊村), Zhengyi Guan (正一观[8]), cradleland , Taoist Temple (道观), legend, Zhang Daoling, Alchemy (炼丹[9]), founder, humanity, shengguan (升棺ascending coffin), dream seeking, shengguan (升官[10]promotion), Holy Land
Natural value (4227/50.88%)	Danxia landform (丹霞地貌), nature, scenery, Elephant Trunk Hill, sightseeing, Luxi River (泸溪河[11]), landscape, Fairy Water Rock (仙水岩), scenic spot, environment, Fairy Rock(仙女岩), Peach Blossom Island (桃花洲), both sides, park, famous mountain
Culture and nature integration value (510/6.22%)	graceful, handsome, beautiful, beautiful scenery, Xiao Guilin, picturesque scenery, good-looking, fairyland (仙境)
Other value (729/8.04%)	tickets, cheap, ticket price, fee, free, cost performance

with the specific classification shown in Table 5.5. The heritage value embedded in the high-frequency words is semantically divided into four categories: cultural value, natural value, cultural and natural integration value, and other values. The statistical results reveal that the words related to cultural value constitute 34.86%, mainly describing the Taoist culture and the cliff-hanging coffins in Longhushan. Words linked to natural value make up 50.88%, predominantly describing the landscape scenery. Upon classifying the words, it is evident that tourists tend to incorporate their personal emotions, experiences and memories into descriptions of Longhushan's natural landscape, as demonstrated by terms like 'fairyland' and 'Xiao Guilin'. Such words comprise 6.22% of the total, reflecting the characteristic integration of nature and culture.

This study employs Gephi to create a visual semantic network for a more comprehensive analysis of the relationships among high-frequency words. In the network analysis graph, each node symbolises a vocabulary word, and its position is determined by the quantity of its connections. A vocabulary word located closer to the center of the semantic network suggests a greater number of direct connections to other words, indicative of richer topics surrounding that word (Li & Zhang, 2020). The connections between high-frequency words represent semantic relationships; the darker and denser the lines, the stronger the semantic connections. The size of each node reflects its degree of association with other nodes – larger nodes signify higher frequencies of co-occurrence with other vocabulary words.

The semantic network of high-frequency words comprises two core words: 'Longhushan' and 'scenery'. 'Longhushan' serves as the main subject of tourists' evaluations and the nucleus of the tourists' discourse system, while 'scenery' underscores the prominent natural value of Longhushan from the tourists' perspective. Terms such as 'Longhushan – Bamboo-Rafting', 'Longhushan – perform', 'Longhushan – plank road', 'scenery – perform', 'scenery – plank road' frequently appear concurrently. This suggests that the natural elements of Longhushan serve not only as the primary carrier of scenic tourism activities but also as the medium for different cultural exchanges and dissemination. Relationship chains such as 'Longhushan – Taoism', 'Taoism – culture', 'Longhushan – Tianshi' and 'Taoism – cradleland' emphasise the importance of Taoist culture as one of Longhushan's significant resources. The term 'scenery' also co-occurs with 'Tianshi' and 'taoism', indicating the general recognition of the fusion of culture and nature (Figure 5.2).

Research Findings: Heritage Value of Longhushan from the Perspective of Tourists

Tourists' comments on Longhushan reveal that they can recognise the natural and cultural values of Longhushan and interpret the fusion of nature and culture based on their personal experiences. Tourists are drawn to Longhushan not only for its scenic natural landscapes with clear waters and majestic mountains, including its globally celebrated Danxia landforms, but also for its notable cultural elements, such as the enigmatic cliff-hanging coffins. In summary, tourists express the value of Longhushan in the following three ways.

Scenic showcase with breathtaking mountains and beautiful water

Table 5.5 and Figure 5.2 show that 50.88% of high-frequency words relate to natural value, with 'scenery' being a key term besides 'Longhushan'. This implies that Longhushan's unique natural landscape is the main tourist attraction. Many comments describe the overall natural beauty of Longhushan, such as 'the low and continuous mountains that are easy to climb, and the red cliffs that present visually appealing shapes' (LHS0003). Noteworthy attractions such as 'Elephant Trunk Hill', 'Fairy Water Rock', 'Fairy Rock' and 'Luxi River' are frequently

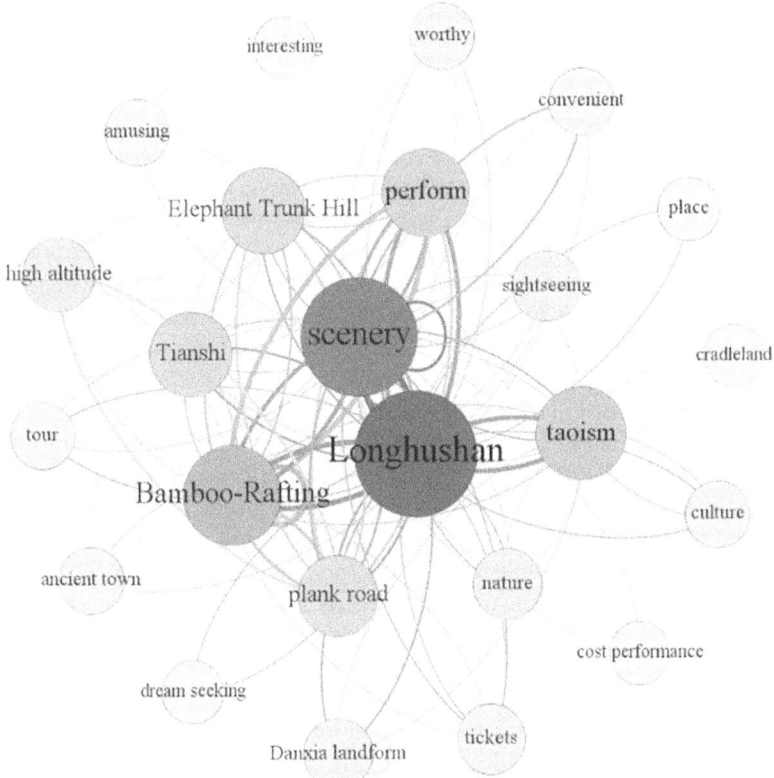

Figure 5.2 Semantic network analysis diagram

mentioned by tourists, with descriptions emphasising the grandeur and spirituality of these landscapes.

'Danxia landform' is another high-frequency word associated with natural value. Influenced by official narratives, tourists are curious about the typical Danxia landforms and express a desire to learn about geology while enjoying the natural scenery. Tourists express that 'Longhushan is a great place to explore geological history and learn about geology' (LHS0118). They can 'travel while simultaneously gaining knowledge of geology' (LHS0118). Being surrounded by the Danxia peaks and forests, they feel a sense of pride because 'the Danxia landform was named in China, and it was primarily promoted by Chinese scholars to become a globally renowned type of scenic landscape' (LHS0245).

Overall, the terms 'mountain', 'rock', 'river' and 'Danxia landform' constitute the primary elements of tourists' perception of Longhushan's natural value. Tourists can directly experience the beauty of Longhushan's mountains and rivers, but their understanding of the scientific value of the 'Danxia landforms' remains at a surface level. Most tourists recognise its title as 'Chinese Danxia' without fully understanding its significance.

Cultural sanctuary: A repository of humanistic beliefs

The profound and extensive Taoist culture is another essential resource of Longhushan. High-frequency terms in Table 5.5, such as 'Upper Pure Ancient Town', 'Upper Pure Palace', 'Zhengyi Guan' and 'Celestial Master's Mansion' indicate tourists' attention to the Taoist culture of Longhushan. Taoism, a religion indigenous to China enriched with profound intellectual wisdom, has made Longhushan, one of the four famous Taoist mountains, a magnet for tourists eager to explore its pivotal Taoist position. Some tourists even expressed that 'the scenery of Longhushan may not be the best among the numerous renowned mountains and rivers in China, but the main attraction here is to understand Taoist culture' (LHS0007). While most tourists have a limited understanding of Taoist culture, they are drawn to Longhushan, known as the 'cradle of Chinese Taoism' (中国道教祖庭) (LHS0004, LHS0005, LHS0017, LHS0068, LHS0144, LHS1026) to 'witness Taoist rituals' (道教法事) (LHS0004, LHS0017) 'immerse themselves in the deep implications of Taoism' (LHS0005, LHS0068, LHS0144) and 'experience the millennium-old Taoist culture, praying for peace' (感受千年道家文化，祈愿平安) (LHS1026). Some tourists, already familiar with some aspects of Taoism before their visit, hope to delve deeper into its mysteries. 'Taoism plays the role of a pharmacy in Chinese culture; its intricacies are worth profound comprehension' (LHS0006), and 'During the Eastern Han Dynasty, Zhang Daoling, the founder of Zhengyi (Orthodox Unity) Taoism, is said to have made an elixir here, and a tiger appeared when the elixir was completed' (丹成而龙虎现) (LHS0001, LHS0005, LHS0074, LHS0689) – such Taoist legends are also attractive to tourists. Additionally, a few Taoist followers consider Longhushan, the birthplace of Taoism, as a global pilgrimage site, dubbed in the tourism industry as the 'top destination for Taoist pilgrimage tours' (LHS0001, LHS0008, LHS0042, LHS0472, LHS1624).

Apart from Taoist culture, Longhushan's cliff tombs and hanging coffins significantly shape tourists' cultural perception. Longhushan is recognised as the cradleland of cliff burial culture in the Southeast Asia and Pacific region. The steep and inaccessible cliff tombs remain an enduring mystery. In 2006, the scenic area used the most primitive lever principle for the performance of ascending coffins, locally referred to as 'shengguan' (coffin-ascending) which phonetically sounds similar to 'shengguan' (promotion). Some tourists specifically come to witness this performance with its symbolic meaning of 'promotion and wealth, good luck and prosperity' (LHS0002, LHS0014, LHS0079). Some even connect the hanging coffin performance with workplace competition and find new meanings in the heritage, stating that 'The process of the coffin-ascending performance is interesting and can be summarised as follows: there are people pulling from above, people blowing from below, and a position in the middle. Originally, 'shengguan' (promotion) and 'shengguan' (coffin-ascending) were completely unrelated matters, but the processes are similar' (LHS0014). This reflects the dynamic nature of heritage, which can be given new interpretations and significance through different social experiences.

Beyond sightseeing, tourists also hope to experience diverse local cultures. The distinctive Taoist culture and cliff tomb culture of Longhushan interact, making the scenic area more attractive. Tourists consider the 'Taoist sacred site, picturesque landscape, and ancient cliff tombs' as the 'three unique attractions of Longhushan' (LHS0017, LHS0063, LHS0129). Having fully recognised the cultural value of Taoist culture and cliff tombs at Longhushan, tourists reinterpret these elements based on current social, cultural and political needs. As a result, the value of these cultural heritage sites continues to evolve and enrich over time.

Spiritual place of emotions and aspirations

Traditional Chinese landscape aesthetics is characterised by the harmonious coexistence of reality and illusion, where the observer becomes one with the observed, and emotions merge into the scenery (Xu et al., 2016). This concept is also evident in tourists' appreciation of Longhushan's aesthetic value. High-frequency descriptors like 'graceful', 'beautiful', 'good-looking' and 'fairyland' reveal that tourists appreciate not only the tangible beauty of Longhushan but also its intangible allure. On the one hand, physical landscapes such as the Danxia landforms and lakes stimulate tourists' sensory experiences of Longhushan, evoking sentiments such as 'the tranquility of the mountains and waters and the peacefulness of the bird songs and flower scents bring us joy and contentment' (LHS0088). On the other hand, the Taoist legends infuse Longhushan with an aura of mystique, as expressed in phrases like 'a place where immortals reside naturally becomes a fairyland' and

> as the saying goes, "the blue water Danxia perched on the tiger and dragon, the cave of heavenly blessings hidden immortal court" (碧水丹霞踞虎龙, 洞天福地隐仙庭). If there is no Longhushan's bells and whistles, it might not have attracted Zhang Daoling. (LHS0009)

Furthermore, tourists often intertwine their personal emotions, experiences and memories into their aesthetic appreciation of Longhushan, fostering an interactive amalgamation of the two. 'Danxia cliffs, strong mountains and rivers, peaks and turns, life is the same' (丹霞峭壁, 气壮山河, 峰回路转, 人生也是如此) (LHS0048) – for this tourist, the value of heritage lies in finding life's meaning. Numerous tourists also employ poetry to portray Longhushan's natural landscapes, such as: 'The stream flows gently, with visible fish when the water is shallow, deep enough for the bottom to be unseen' (水缓时款款而行，水浅处游鱼可数，水深处碧不见底). Accompanied by the mountain rocks, it forms a magnificent scene of 'a stream merging with the glass-like, and countless layers of purple and green hills' (一条涧水琉璃合，万叠云山紫翠堆) (LHS0055).

This poetic verse was penned by Zhang Jixian (张继先[12]), the 30th generation celestial master of the Zhengyi Guan, during his stay at Longhushan. Immersed in the actual scenery, tourists seem to relive what Zhang Celestial Master once observed and felt.

> The small boat goes against the current, and the clear Lu River reveals the pebbles at the bottom. Ignoring my longing gaze, they flow smoothly backward together with the reflected peaks in the water. Truly, it creates the artistic conception of "people walking in a mirror, birds flying in a windless screen" (人行明镜中，鸟度屏风里) (LHS0034)

Drifting on a bamboo raft in the gorge, one can almost sense the same ambience as Li Bai's visit to Qingxi (青溪). Through vivid language, tourists articulate the immersive aesthetic experience of Longhushan, bridging the gap between the observer and the observed and establishing a profound connection between people, nature and culture. The aesthetic value of Longhushan is assigned by human perception, and tourists incorporate elements of nature, culture and personal memories and experiences into their interpretation of Longhushan's aesthetic value, thereby enriching and diversifying the perception of this heritage.

Discussion

Using Longhushan as the research subject, this study discovers that tourists' understanding of heritage value is multifaceted. Initially, tourists appreciate the natural landscape of Longhushan's Danxia landform, aligning with its highly recognised natural value in official discourse. Additionally, Longhushan's Taoist culture and cliff tomb culture are significant attractions for tourists. They are captivated by the cultural pride associated with Taoism and the intrigue of the cliffside coffins. Over time, historical and cultural elements have been imbued with new meanings. Also, in recognising the aesthetic value of Longhushan, tourists incorporate personal emotions, memories and experiences, reflecting the value of the integrated cultural and natural heritage site.

The study identifies variations in the construction of heritage value between three stakeholder categories: the OUV of World Heritage, domestic scholars' research and tourist discourse. Firstly, the OUV of World Heritage tends to be singular, focusing on the unique attributes of the heritage site on a global scale, and its significant contributions to societal civilisation and human progress. Being part of the 'China Danxia' series of heritage sites, Longhushan's OUV is primarily explicated in terms of its natural value from aesthetic and scientific perspectives. Both domestic scholars and tourists concur on the interpretation of Longhushan's natural value through the OUV concept. Consequently, domestic scholars incorporate Longhushan's aesthetic value into its natural value (Table 5.2) and comprehensively highlight Longhushan's exceptional contributions from aesthetic, scientific and biodiversity perspectives. Although tourists might lack an in-depth understanding of the scientific value concerning geology, landforms and biodiversity, they exhibit a characteristic of integrating emotions into understanding Longhushan's aesthetic value. Unlike the AHD represented by the OUV and domestic scholars, tourists articulate their personal experiences and emotions vividly and subtly, incorporating them into their aesthetic interpretations of Longhushan. Therefore, the aesthetic value from the perspective of tourists extends beyond mere natural value and reflects the integration of natural and cultural values advocated by the Fuzhou Declaration.

Both Chinese scholars and tourists recognise the cultural significance of Longhushan, but their perspectives vary. Scholars primarily focus on the inherent Taoist culture, drawing attention to the historical and cultural values of elements such as the 'mystery of cliff tombs', 'sacred Taoist sites' (洞天福地[13]), and Longhushan as the 'cradleland of Taoism'. In contrast, tourists highlight the national pride elicited by Taoism as a native Chinese religion. Additionally, tourist discourse exhibits greater diversity, interpreting the symbolism of 'shengguan' (coffin-ascending) as 'shengguan' (promotion), and drawing a parallel between the 'process of ascending coffins' and the 'process of getting promoted'. This reflection mirrors the unique cultural curiosity intrinsic to Chinese folk culture and enriches the heritage with deeper cultural connotations.

This study, through a comparative analysis of the OUV of World Heritage, research by domestic scholars, and tourist discourse, posits that during the process of World Heritage nomination, Chinese experts formulate the official narrative of Longhushan based on the Guidelines' requirements and the assessment opinions of the IUCN, encapsulating the OUV of World Heritage. Domestic scholars then augment this narrative by elucidating further on Longhushan's natural and certain aspects of its cultural value within the World Heritage value framework, drawing from the OUV. Tourists, albeit influenced by the official narrative, interpret Longhushan's value through a lens of rich personal emotions, romanticism and an ability to explore the intertwined cultural and natural values of Longhushan that may have been underrepresented in the official discourse.

Against the backdrop of the flourishing cultural tourism industry, the profound understanding of heritage by the public, represented by tourists, provides a more enriching interpretation of heritage sites. Through the public's perception of nature, culture and aesthetics, management can have a clearer grasp of the unique aspects of the heritage site, explore cultural resources within the area, and consequently formulate more targeted conservation strategies. This not only helps balance the relationship between the development of cultural tourism and the preservation of heritage sites but also provides a more diverse and attractive promotional perspective for the heritage site, contributing to its broader appeal. Thus, during the declaration and management of heritage sites, it is essential to consider public discourse. This can be accomplished by recording the perceptions and requirements of various stakeholders, including domestic experts and scholars, tourists, and community residents, through online discourse analysis, interviews and questionnaires. This can help encapsulate the value of heritage as constructed by public discourse. Concurrently, big data can be leveraged to gather people's poetic aesthetic descriptions of the heritage, integrating their personal memories and experiences to understand the heritage's meaning better. This approach can enrich the heritage's diversified values, foster an interactive connection between people and the heritage and ultimately bring the heritage 'to life'.

Drawing on Matteucci *et al.*'s three different visions of the future development of cultural tourism, the case of Longhushan exemplifies both utopian and heterotopian visions of cultural tourism due to its integration of multiple, contrasting perspectives on heritage value within a broader tourism framework. Unlike the dystopian view that is characterised by degradation and commodification, this case reflects a utopian ideal in which diverse stakeholder viewpoints coexist. The OUV, local scholar interpretations, and tourists' personal experiences all contribute to a richer, multi-faceted understanding of Longhushan. By highlighting the disparities between authoritative heritage discourse and public perception, the case also reflects the heterotopian principle of creating spaces where alternative, resistant or emergent narratives can challenge or complement dominant ones. It also emphasises the necessity of public participation in cultural tourism management which should help foster creative, socially inclusive and culturally enriched experiences within the existing tourism system. In a nutshell, this chapter aims to provide both theoretical and empirical foundations for transforming cultural tourism practices and offer practical pathways for developing more resilient and inclusive, utopian and heterotopian visions of cultural tourism in the future.

Conclusion

In envisioning the future of cultural tourism, particularly through the lens of World Heritage Sites like Longhushan, it becomes imperative to recognise the evolving dynamics between heritage interpretation and the stakeholders involved. The case of Longhushan exemplifies a critical juncture where the acknowledgment of a site's comprehensive value – both natural and cultural – is paramount for fostering an inclusive understanding of heritage. This understanding, as demonstrated through tourist discourse, not only enriches the narrative surrounding a site but also aligns with the broader objectives of sustainable tourism, emphasising the need for a participatory approach in heritage management.

The transition towards a more bottom-up approach in heritage interpretation and management is instrumental in addressing the multifaceted values that tourists associate with sites like Longhushan. Such an approach necessitates a shift in perspective, recognising tourists not just as passive observers but as active participants in the heritage discourse. This shift is crucial in an era where the interplay between technology and social media platforms has democratised information sharing, enabling tourists to contribute significantly to the narrative of heritage sites. Furthermore, acknowledging the voices of tourists and incorporating their insights into heritage management can enhance the authenticity and relevance of the heritage experience, thereby fostering a deeper connection between the site and its visitors.

The integration of bottom-up approaches in heritage interpretation also aligns with global trends towards more sustainable cultural tourism practices. As tourists increasingly seek meaningful and immersive experiences, understanding and valuing their perspectives can lead to more effective conservation strategies that balance visitor needs with the preservation of the site's OUV. Moreover, this approach can serve as a catalyst for engaging local communities, ensuring that the benefits of tourism are equitably distributed and that heritage sites continue to hold significance for both local and global audiences.

However, the implementation of such participatory approaches faces challenges, including the need for institutional support, policy frameworks that prioritise stakeholder engagement and mechanisms for effectively capturing and integrating public discourse into heritage management. Overcoming these challenges requires concerted efforts from all stakeholders, including heritage bodies, policymakers, scholars and the communities connected to these sites. By fostering collaborative dialogues and leveraging technological advancements, the future of cultural tourism can be reshaped to be more inclusive, sustainable and reflective of the diverse values that heritage sites like Longhushan embody.

In conclusion, the exploration of Longhushan's heritage value through the lens of tourist discourse underscores the critical need for a more nuanced and inclusive approach to heritage interpretation and management. Embracing a bottom-up methodology not only enriches the heritage narrative but also aligns with the evolving expectations of tourists and the broader objectives of sustainable cultural tourism. As we move forward, the insights gleaned from such discourses can inform strategies that ensure the enduring relevance and conservation of World Heritage Sites, ultimately contributing to a more vibrant and inclusive vision for cultural tourism.

Notes

(1) ROST-CM6 can perform word segmentation, word frequency statistics, part-of-speech sentiment analysis and vocabulary co-occurrence analysis on text information.
(2) Gephi is a JVM-based complex network analysis software, mainly used for visual analysis of various networks and complex systems, dynamic and hierarchical graphs.
(3) Zhang Daoling, the founder of Taoism. He was also known as Zhang Tianshi (张天师) because of his initial founding of Wu Dou Mi Dao (五斗米道), also known as Tian Shi Dao (天师道).
(4) The landscape of Guilin is the representative of China's landscape, enjoying the reputation of 'Guilin landscape is the best in the world'. The scenery of Longhushan is similar to that of Guilin, and tourists to Longhushan can't help but think of Guilin and call it 'Little Guilin'.
(5) Upper Pure Ancient Town belongs to the scenic area of Longhushan and has a history of thousands of years. It is known as the first Taoist town in China for its profound Taoist culture.
(6) Upper Pure Palace is a famous Taoist temple, located in the scenic area of Longhushan. It is the place where Zhang Tianshi carried out Taoist activities.
(7) Tianshi refers to Zhang Daoling, the founder of Taoism, and his disciples in the mantle, the head of Taoism.
(8) Zhengyi Duan is a symbol of the ancestral court, where Zhang Daoling used to make alchemy in the Eastern Han Dynasty.
(9) Taoists use cinnabar to make medicine that makes people immortal.
(10) Ascending coffin is a performance form of traditional culture in Longhushan Scenic spot. Promotion is the pursuit of people in the workplace, both of which have the same pronunciation in Chinese.
(11) The Luxi River is a tributary of the Xin River and passes through Longhushan.
(12) The 30th generation of Tianshi Dao, a famous Taoist at the end of the Northern Song Dynasty.
(13) Famous mountains and scenic spots inhabited by Taoist gods.

References

Chen, M. (2009) Research about religious buildings and cultural in Jiangxi province – Taoist buildings. *Sichuan Building Science* 35 (6), 298–300.
du Cros, H. and McKercher, B. (2015) *Cultural Tourism*. Routledge.
Fu, J. and Yu, W. (2019) Exploration of the utilization and development of tourism resources in Longhushan scenic area. *Guangdong Canye* 53 (8), 93–94.
Guo, F., Li, X. and Jiang, Y. (2012) *Danxia Landform and Tourism Development in Mount Longhu*. Geological Publishing House.
Guo, F., Zhu, Z., Huang, B. and Jiang, Y. (2013) Cretaceous sedimentary system and its relationship with Danxia geomorphology in Xinjiang basin, Jiangxi province. *Acta Sedimentologica Sinica* 31 (6), 954–964.
Harrison, D. (2021) *Tourism, Tradition and Culture: A Reflection on their Role in Development*. CAB International.
Huang, J. (2010) *Danxia Landform*. Science Press.
ICOMOS (1994) *The Nara Document on Authenticity*. See http://iicc.org.cn/Publicity1Detail.aspx?aid=926 (accessed June 2022).
Jiang, X., Pan, Z., Xu, J., Li, X., Xie, G. and Xiao, Z. (2006) Late cretaceous eolian dunes and wind directions in Xinjiang basin, Jiangxi Province, China. *Geological Bulletin of China* 7, 833–838.
Li, R. and Zhang, H. (2020) Research on tourism destination image perception based on text mining – a case study of Korean tourists' evaluation of Qingdao. *China Tourism Review* 1, 41–53.

Li, W. and Chen, G. (2010) China Danxia: Four years of arduous journey for world heritage nomination. *Hunan Daily*. See https://hnrb.voc.com.cn/hnrb_epaper/page/1/2010-08/11/04/2010081104_pdf.pdf (accessed May 2020).

Liao, S. (2019) Characteristics and conservation of shallow craquelure landscapes in the Longhushan area, Jiangxi province. *Gansu Science and Technology* 35 (20), 30–31.

Liuzza, C. and Meskell, L. (2021) Power, persuasion and preservation: Exacting times in the world heritage committee. *Territory Politics Governance* 11 (77), 1265–1280.

Matteucci, X., Koens, K., Calvi, L. and Moretti, S. (2022) Envisioning the futures of cultural tourism. *Futures* 142, 103013.

Meskell, L., Liuzza, C., Bertacchini, E. and Saccone, D. (2015) Multilateralism and UNESCO world heritage: Decision-making, states parties and political processes. *International Journal of Heritage Studies* 21 (5), 423–440.

Smith, L. (2006) *Uses of Heritage*. Routledge.

Smith, L. (2015) Intangible heritage: A challenge to the authorised heritage discourse? *Revista d'Etnologia de Catalunya* 40, 133–142.

Smith, L. (2020) *Emotional Heritage: Visitor Engagement at Museums and Heritage Sites*. Routledge.

Tang, L. and Wan, J. (2010) Nomination for heritage of danxia in china: A story of death and death. *China Newsweek* 36, 76–80.

UNESCO (2010) *China Danxia*. See https://whc.unesco.org/en/list/1335 (accessed May 2022).

UNESCO (2021a) *Operational Guidelines for the Implementation of the World Heritage Convention*. See https://whc.unesco.org/en/guidelines/ (accessed June 2023).

UNESCO (2021b) *Fuzhou Declaration*. See https://whc.unesco.org/document/188530 (accessed 6 June 2023).

Winter, T. (2015) Heritage diplomacy. *International Journal of Heritage Studies* 21 (10), 997–1015.

Xiaoqian, H., Li, C. and Xu, J. (2018) The features of tourists perception of popular science education in Longhushan global Geopark. *Journal of Arid Land Resources and Environment* 32 (8), 202–208.

Xu, X., Yang, R. and Zhuang, Y. (2016) Aesthetic evaluation framework research of Chinese famous mountainous scenic sites. *Chinese Landscape Architecture* 32 (9), 63–70.

Yao, C. (2020) Discussion on the external artistic features of the music of tianshi dao rituals in Longhushan, Jiangxi. *Home Drama* 8, 52–54.

Yu, J. and Zhang, C. (2020) Review on heritage discourse. *Study On Natural and Cultural Heritage* 5 (1), 18–26.

Zeng, Z., Zhang, D., Lin, R., Chen, L., He, T., Ye, J. and Zheng, Y. (2022) Research on tourist destination imagination and perceptual characteristics based on web text analysis – taking Yangzhou Geyuan as an example. *Journal of Southwest University (Natural Science Edition)* 44 (1), 194–201.

Zhang, R. (2020) *Chinese Heritage Sites and their Audiences: The Power of the Past*. Routledge.

Zhang, R. and Brown, S. (2022) Benefit or burden the world heritage listing of Libo Karst China. *International Journal of Heritage Studies* 28 (5), 578–596.

Zhang, R. and Brown, S. (2023) Comparing landscape values and heritage stakeholders: A case study of West Lake cultural landscape of Hangzhou, China. *International Journal of Cultural Policy* 29 (2), 184–201.

Zhang, Y. and Guan, X. (2022) The transformation of rural dwellings from the perspective of tourism economy: A case study of the Mosquito-Free village in Longhushan, Yingtan City. *Art Education Research* 8, 87–88.

Zhu, Z., Huang, B., Guo, F., Zheng, H. and Jiang, Y. (2012) Cretaceous braided river facies sediments and Danxia landform development characteristics in Longhushan world geopark, Jiangxi. *Acta Geoscientica Sinica* 33 (3), 379–387.

Part 2

Consumption

6 Edifying Slow Cultural Tourism Concepts and Practices

Jelena Farkić

Introduction

Anthropocene, a geological epoch shaped by human activities that is marked by significant human impact on the Earth's geology and ecosystems, is closely related to the considerations in envisioning the future of tourism (Holden et al., 2022). Scholars have particularly interrogated the influence of capitalism and profit-centred development, prompting critical reflections on how economic structures shape environmental transformations, while questioning the consequences of industrial expansion, urbanisation and the commodification of global cultural landscapes. The anthropogenic influence, driven by economic imperatives, often jeopardises cultural heritage, cultural artefacts, historic sites or indigenous knowledge (Bui et al., 2020). This not only requires understanding the tangible alterations to our planet but also the intangible erosion of cultural identities and narratives, emphasising the urgent need for holistic and sustainable approaches to the development of cultural tourism.

Rantala et al. (2020) called for alternative imaginings, conceptualisations and practices of tourism for the ongoing Earthly crisis. Cultural tourism has been traditionally associated with human-created landscapes, cultural and historical sites and communities and the promotion and preservation of both the tangible and intangible expressions of human culture. However, a recognition of the intertwined relationship between nature and culture, prompting a rethinking of the nature/culture binary, has challenged the notion that cultural tourism is solely confined to human-made environments (Harrison, 2015; Hartley, 2016). Natural landscapes often play a pivotal role in cultural tourism, serving as the backdrop for cultural sites or events. They are indeed an integral component of the cultural tourism experience through which tourists immerse themselves into the legends, rituals, values or traditions embedded in and shaped by those landscapes (Hall et al., 2015).

Cultural tourism therefore in many ways plays a crucial role in enabling deeper connections to and cultivating an appreciation for natural knowledge and values embedded in diverse, and emergent, cultures globally, potentially blurring the boundaries between human and natural communities. Such understanding is mainly related to, for example, indigenous epistemologies and various cultural knowledge systems that refrain from regarding nature as a mere object for detached utilisation or scientific scrutiny (Monaheng, 2016; Prasetyo *et al.*, 2023). To this end, Greenwood (2013) urges that we have to begin to qualify our cultural and economic goals with socioecological thinking, by asking what kind of growth and development will serve diverse people, places, species and cultures, now and in the long-run. Consequently, this study seeks to join the scholarly debates that aim to decolonise and disrupt conventional narratives of cultural tourism in the Anthropocene and explore the potential of developing alternative relationships among humans and their environments (Hockert *et al.*, 2023). The study, therefore, builds on these ideas and moves beyond the dominant spaces of culture, while through the feminist new materialist lens, it aims to explore the alternative places, narratives and dynamics that are emergent within *natureculture* realms, and their possibilities for slow cultural tourism development.

Situating Slow Cultural Tourism within Earthly Crisis

In light of the destructive impact of the Anthropocene and the challenges affecting cultural heritage and tourism dynamics, it becomes imperative to adopt more responsible tourism practices. This requires a thorough examination of how destination managers interpret culture and incorporate it into the development of tourism products (Matteucci *et al.*, 2022a). Additionally, the policy decisions made by destinations hold consequences for regional ecosystems and the well-being of local residents. The success of destinations relies on their capacity to instigate positive changes that may improve the overall quality of life for local communities. Higgins-Desbiolles *et al.* (2019) advocate for a redefinition of tourism, to prioritise the rights and benefits of local communities over those of tourists and profit-making.

To address the concerns related to tourism governance, Calvi *et al.* (2020) put forth a model outlining four potential scenarios for cultural tourism. The model distinguishes between two overarching governance approaches – economy-oriented and community-oriented – considering the demand for cultural tourism. The former leverages local cultural resources to maximise economic benefits for the tourism industry, while the latter seeks to revitalise local cultures and enhance community well-being. One of the proposed scenarios, which was subsequently developed in Matteucci *et al.*'s (2022a) study, that of community-driven slow cultural tourism, emerges from the intersection of rising demand for slow(er) tourism experiences more broadly. The emergent niche markets attract individuals seeking the 'authenticity' of local culture and are willing to pay higher prices for such experiences, in contrast to fast cultural tourism demand desiring standardised and less authentic experiences. In the context of community-driven slow cultural tourism, participatory initiatives emphasise the community's definition of 'local culture' and its social and ecological dimensions. Cultural offers centre mainly around local and regional products, utilising traditional production methods and employing local human capital. Mass-produced items are eschewed to ensure personalised tourist experiences, albeit at relatively higher prices. While innovative strategies may be needed to attract specific cultural tourists, the authors caution that niche markets may not generate sufficient benefits for community prosperity, posing potential risks for destinations.

The above interpretation of community-driven slow cultural tourism in many ways aligns with feminist, postcolonial and indigenous perspectives on developing tourism more locally, and more ecologically. It also follows the more recent scholarly efforts to explore diverse forms of interactions, offering a novel lens that acknowledges the complex entanglements of human and non-human agents within cultural landscapes (Grimwood *et al.*, 2019; Höckert *et al.*, 2023). Such an approach embraces environmental ethics, rooted in a worldview that extends relationships to encompass all materials, things and beings, highlighting reciprocity and responsibility towards others, including the land itself (Boukhris, 2020). In cultural tourism settings, crucial is the process of developing authentic and meaningful relationships, those that transcend mere economic interests, whether with the land and nonhuman nature, local community members or tourists. This creates space for and enables deeper and more meaningful encounters, pro-environmental behaviours and sensitising cultural experiences, as discussed by Saari *et al.* (2020) within the context of Sámi culture. Indeed, decolonising cultural tourism and developing cultural sensitivity allows us to embrace different perspectives, places and forms of interactions (Viken *et al.*, 2021). Accounting for the agency of the environment in which tourism activity takes place might therefore open up new avenues for thinking, doing and researching cultural tourism.

The concept of slow cultural tourism shares some common principles with the broader philosophy of slow tourism (Caffyn, 2012; Fullagar *et al.*, 2012). Essentially, they share core values of authenticity, community involvement, and a deliberate, responsible approach to travel. Emphasis is placed on a decelerated pace of movement, quality over quantity and facilitation of deeper and more meaningful experiences. Similarly, slow tourism encourages proximity and hyperlocal experiences and avoids rushed itineraries and mass-produced experiences, as discussed by the proponents of the idea of slow cultural tourism (Calvi *et al.*, 2020). It supports the community engagement and sustainability focus often associated with slow tourism (Oh *et al.*, 2016). As explicated by Matteucci *et al.* (2022a), slow cultural tourism, involves a gradual exploration of specific 'authentic' elements of the local culture whose meaning is emergent and shifting, and generated by the local community, whether through plants, foods, small-scale festivals or arts and crafts. Central to the product, however, is the local community and its agency in generating meaning, creating local narratives, and storying the cultural landscapes while at the same time inscribing the local, tacit knowledges, traditions, values and beliefs in those discourses.

Taking into account the eco (rather than ego)-centric orientation of slow cultural tourism, we can also draw similarities with slow adventure, a subset of adventure tourism (Varley & Semple, 2015). This type of travel encourages a more deliberate and reflective approach to outdoor experiences, prioritising engagement with the natural environment and local communities. Similarly, cultural exchange among individuals from different cultural backgrounds is crucial (Wengel, 2021). The slow pace provides ample opportunities for meaningful encounters. The cultural and social exchange deepens the experience and contributes to authenticity as tourists learn about local customs, traditions and ways of life directly from the local communities and local, 'homegrown' guides. Farkić *et al.* (2020) have explored how the quality of slowness allows tourists to flourish through savouring the experiences in wild nature. This is linked to positive psychological outcomes of well-being and mindfulness thus greatly contributing to social sustainability. The emphasis on sustainability more broadly has prompted adventure tourism providers to promote eco-conscious behaviour, and appreciation for environmental conservation and ethical conduct through product design (Beames *et al.*, 2022).

In situating community-oriented slow cultural tourism within broader academic discourse, it is important to briefly attend to local community agency. By way of example, Matarrita-Cascante *et al.* (2010) characterise community agency as the establishment of local connections that enhance the adaptive capabilities of individuals within a shared locality. Recognition of shared local values and traditions is vital for fostering community agency (Guri *et al.*, 2021). At the core of the concept of community agency is the cultivation of communities' abilities to access and utilise resources, not only to improve their well-being but also to foster their joint commitment to community development through sustainable practices. To this end, the community plays a pivotal role in facilitating engagement in initiatives that support cultural tourism, including enabling access to business and entrepreneurial opportunities, from which the broader community will benefit (Su & Wall, 2015). In slow cultural tourism, crucial are the elements such as social and human capital, as well as an increased awareness of local identity; this all being linked to community agency and the ways in which it emerges, forms and is negotiated within specific place and time (Mzembe *et al.*, 2023). Moving beyond strengthening collaboration and sharing the benefits of cultural tourism, the community aims to actively preserve both natural and cultural resources which are intended to secure an equitable distribution of economic and social benefits, and enhance both community and environmental wellbeing.

Harking back to the Anthropocene and its challenges, destinations grapple with policy choices that carry significant repercussions for regional ecosystems and the well-being of their resident population. Evidently, the prosperity of destinations hinges on their ability to bring about positive transformations that broadly enhance the welfare of local communities. Scholars urge for an urgent paradigm shift in policy to secure a more sustainable future for all living beings. By way of example, Matteucci *et al.* (2022b) propose that tourism destinations follow a new materialist governance approach, which prioritises the needs and concerns of local communities over those of tourists and the travel trade, and are more environmentally sensitive. While there is a possibility of ongoing change, scholars express concern that this transformation might be gradual and initially manifest as isolated pockets of resistance

In discussing how the response to changes might be slow and manifest itself initially as 'pockets of resistance' and 'bubbles of ethical consumption', Matteucci *et al.* (2022a) draw on Michel Foucault's (1986) notion of heterotopia, offering useful perspectives on societal structures and community dynamics, while approaching them as unconventional spaces or counter-sites that deviate from societal norms and can exist in various forms across cultures. They embody processes of alternate social ordering, allowing people to use them in diverse ways. Despite being distinct spaces, heterotopias maintain connections to other sites, enabling free entry and exit, and they serve a function in relation to the broader social landscape, offering an illusion of utopia or acting as sites of transgression where normative behavioural patterns can be escaped.

Matteucci *et al.* (2022a) engaged with the concept of heterotopia as a useful lens to discuss the resistance to mainstream neoliberal practices by introducing us to the notion of pockets of resistance within tourism. The authors propose that heterotopias can emancipate their inhabitants, providing a sense of liberation or freedom. These spaces, situated outside conventional societal structures, may feature distinct rules, meanings or functions. Further, the authors suggest that heterotopian communities are based on the precepts of solidarity, social justice, cultural diversity and slow encounters. The value of such pockets of resistance lies in facilitating the interaction of diverse stakeholders, including nature, which allows for the exchange of ideas, collaboration and the establishment of new networks. In heterotopias, creative activities are key, as they also facilitate multicultural and multispecies understanding, diversity, and above all, foster environmental care. Heterotopias, with their emergent qualities, might therefore offer a productive framework through which to explore the emergent cultural tourism models across various contexts.

Thinking with New Materialism

In this chapter, I engage with post-qualitative inquiry and feminist new materialist theories to explore the voices of members of what will be termed as heterotopian communities within the broader context of slow cultural tourism. Rather than being primarily concerned with our legitimacy as qualitative researchers and adhering to rigorous methodological approaches, I stay with the unsettling qualities that our 'difference' can produce (Fullagar, 2017). Here, I embrace post-qualitative inquiry through 'reading' and engaging with theory and empirical material in a somewhat unconventional way (Fox & Alldred, 2016). Rather than laying out the methodological approach, I allowed myself to maintain messiness while engaging with the methodology of 'getting lost' in theory and real-life situations (Lather, 2016). Indeed, post-qualitative research does not attempt to operate from a perspective of critical objectivity but rather acknowledges the situated, partial, ethical, relational, posthuman and responsive ways of knowing that have been developed in feminist studies.

Post-qualitative inquiry, however, is immanent. As St. Pierre (2019) explains, it never exists, it never is, it is always in becoming. It emphasises the inherent presence and emergence of meaning within a particular context, which helps redefine the research process. This approach aligns with the broader trends in contemporary philosophy, including new materialism, new empiricism or posthumanism, which also incorporate the concept of immanence, while challenging traditional notions of subjectivity, agency and the boundaries between humans and non-humans. It does not seek to outright reject or negate these structures but rather aims to dismantle them in a way that opens up new possibilities, formulations and perspectives.

New materialism shifts away from privileging culture and instead centres on what Haraway (2013) terms 'naturecultures'. Feminist new materialism is transdisciplinary, traversing disciplines such as philosophy, cultural studies, environmental studies and social theory, offering a more relational and process-oriented lens for analysing the complexities of our entangled existence. It embraces a monist view of the human, rejecting the anthropocentrism and dualisms prevalent in humanities while placing an emphasis on matter (Barad, 2007; Braidotti, 2013). The interconnectedness of human and non-human elements are considered as co-constitutive forces. New materialism therefore redefines the relationship between matter and meaning and emphasises the agency, vibrancy and vitality of non-human entities in shaping the world (Rantala *et al.*, 2020).

The feminist new materialist perspective holds relevance to the realm of slow cultural tourism, offering a lens through which to explore and understand the relationships between humans and non-humans. Thus, taking this approach allows us to reconceptualise our understanding of reality and pay attention to non-human entities such as landscapes, artefacts and ecosystems in the meaning-making processes. This, in turn, allows for a more holistic understanding of culture, to move beyond human-centric narratives to recognise the emergent meanings and relationships impacted by the agency of matter, things and other species, and explore the possibilities for future thinking about slow cultural tourism.

Attuning to Šumadijan Heterotopian Community

> A local legend goes: 'Everywhere there were forests, settlers called relatives to come cut down the woods and occupy the land as much as they wanted; it was so impassable, that one could walk for days through it, without seeing the sun...'

To explore the practices of communities, I made a serendipitous virtual journey to the centre of my home country, Serbia. Šumadija region is located at the heart of the country, receiving its name after dense forests (*šuma* means forest in Serbian) that covered the region in the past, particularly in the 16th and 17th centuries. Šumadija offers a highly conducive environment for the development of tourism, particularly in rural areas. All the villages in this region feature attractive landscapes and authentic traditional rural homes, complemented by natural and man-made attractions. The region boasts well-preserved traditional architecture, farm layouts, ancient crafts, local customs, traditional cuisine and the warm hospitality of the residents (Mandarić *et al.*, 2017). Alongside various prospects, however, challenges impede the progress of tourism development in the region, such as agricultural restructuring, diminishing service offerings, depopulation, counter-urbanisation, communication and infrastructure deficiencies and environmental degradation. To preserve and promote local identity and tradition, there have been numerous entrepreneurial initiatives of the local population that contribute to tourism development by providing opportunities for engagement in activities such as camping, cycling, ethno-wellness, bushcraft or cultivating local organic produce.

During my involvement in a tourism related project, I inadvertently discovered a community of individuals living in and offering tourism experiences in Šumadija. Through referrals, I connected with people who shared similar lifestyles and engaged in similar tourism activities within the region. Over several conversations we had online, I realised that this community was tightly-knit, defined by its geographic location, shared values and ways of life. What unites them is the realisation that a capitalocentric lifestyle is not sustainable in the long term. They have chosen to 'go back to basics', offering small-scale tourism activities and embracing a different way of life away from the hustle and bustle of urban environments and the imposed fast pace of life. Notably, their cohesion was rooted in a robust 'word of mouth' network, where mutual recommendations strengthened their local identity and communal spirit. It became evident that these individuals forged a distinctive Šumadijan community by emancipating themselves from and resisting the capitalist system. They built their community on principles of solidarity, mutual support, engaging in slow activities and fostering care.

Engaging in post-anthropocentric theorising enabled me to gain deeper insights into the relational dynamics of one such heterotopian community, and understand the significance of open-mindedness, messiness and flexibility in research. Critical was the spontaneous, anecdotal and unintentional nature of the 'data collection' through online conversations, phone calls, social media messages exchange, or participation in open competition to support their local initiatives. Parts of the conversations are presented across the ensuing pages. They are discussed in relation to the theory I have been working with to narrate the story of the possibilities of (slow cultural) tourism taking place within one emergent heterotopian community.

Pockets of Resistance

The below quotes usefully illustrate Matteucci *et al.*'s (2022a: 6) proposition that 'a heterotopian future of cultural tourism would manifest itself as the multiplication of bubbles of ethical consumption and practices within a slowly decaying mainstream neoliberal order'. Marko's reflection helps us understand how the alternative visions and possibilities come together, through the rejection of the meaninglessness of modern life and embracing 'new' ways of being, together with others:

> Recognizing the flaws in conventional education, I decided to return to the countryside, the place I had left three decades ago. Here, in our community, we've formed a network and initiated collaborative efforts. What unites us is a shared aspiration for freedom, each of us defining and experiencing it in our unique ways. The shift is not just philosophical; it induces a tangible change in our body chemistry. I find it perplexing how some endure the pressures of a system that materializes success as owning a car and going on frequent trips. The entire system seems rigged, producing unfavourable outcomes. Something rooted in a bad intention cannot yield positive results. Capitalism, as it stands today, was conceived with intentions that seem inherently sinister; we exist for it rather than it serving us. I've realized that it will never provide me with what it takes away, and I refuse to be complicit in that narrative.

Marko's emphasis on freedom was not merely a philosophical concept but was seen as a tangible transformation in body and lifestyle. His critique extends to societal norms that equate success with material possessions and frequent travel, viewing the entire capitalist system as flawed, thus affecting his decision to 'return to the countryside'. Capitalism, in its present form, prioritises its interests over the well-being of individuals. Choosing to live in a rural setting with his family and domestic animals, Marko embodies a shift away from conventional urban lifestyles and embraces a more interconnected existence with natural environment. While challenging the dominant narratives associated with urban-centric and materialistic ideologies, Marko emphasises the idea that living in harmony with the environment can contribute to a more sustainable and fulfilling life. Filip, the owner of the campsite located at the hillsides, reported something similar, in a somewhat Weberian way:

> I had a secure and well-paid job in Belgrade, along with my own NGO project that brought in good income. Yet, caught in the routine of home, work, and back, the repetitive cycle made me reconsider. As an economist, I realized I was spending around 25 days a year just on commuting, which, in my thirties, felt too precious to waste. That realization prompted me to resign from my job and try my luck in the field of tourism. [...] So, it's my escape from the harsh realities of life, allowing me to do something I love, to be surrounded by like-minded people, and the fact that I can also make money from it is a bonus.

The quote shows the resistance toward the standardisation imposed by mainstream culture. The routinised pattern of life and the realisation of spending a significant portion of the year on this repetitive cycle prompted Filip to rethink priorities. The decision to resign and venture into tourism signifies a deliberate escape from the perceived harsh realities of life. The motivation is not solely financial; it is about engaging in meaningful pursuits, being surrounded by like-minded individuals, and finding personal satisfaction. Here, we begin to see the emergent heterotopian community, represented as a space where individuals cross paths with like-minded others, and where they can thrive in tune with their values and passions.

It was also Kosta, an avid outdoor enthusiast with a diverse range of interests, including camping, mountaineering, rock climbing, alpinism, horseback riding and archery, who shared similar views. Kosta hosts bushcraft courses while embodying a lifestyle that intertwines his outdoor activities with a deep connection to local culture and traditions. His unconventional approach to camping, opting for a tarp instead of a tent and occasionally sleeping by the fire, reflects a deliberate choice that aligns with embodied outdoor experiences and slow adventures (Varley & Semple, 2015). Through integrating local practices into his outdoor pursuits, Kosta has transformed his activities into an authentic lifestyle that goes beyond mere recreation, blurring the boundaries between nature and culture. He nostalgically explains:

> Well, you see, I think I could call myself old-fashioned. I prefer the old days when people were one with nature. You know, when you were not

using a car, you were riding a horse, you weren't using plastic, you were making your stuff from leather or wood. Yeah, I feel like I'm where I'm supposed to be, in my natural habitat. So when I go to the city, sure, I'm a normal person. I can work in the city, but it's choking me, you know? It's suffocating. And then when I'm out, I feel free, I can breathe. I can do what I enjoy to do. I don't get paid a lot for what I'm doing, but at least I do what I love, I choose to live modestly and happily, rather than working a job I know where I can earn a lot of money but be nervous all the time.

This quote resonates with new materialist ideas in emphasising a holistic and interconnected relationship with the environment. Kosta expresses a preference for the 'old days' when people were more attuned to nature, highlighting practices like riding horses, avoiding plastic and crafting items from natural materials. The quote recognises the agency of non-human entities (nature, horses, wood) for overall well-being while rejecting the dominant materialistic and environmentally degrading practices. The desire to live modestly and happily, prioritising personal fulfilment over monetary gain, reflects a commitment to alternative values associated with heterotopian ideals.

Creative Entanglements

In constructing heterotopias, the interactions and relationships within the local community emerge as paramount. Heterotopias as spaces of creativity, social relationships and knowledge envisioned by Matteucci *et al.* (2022a) inherently rely on the active involvement and contribution of the community members. As the below quotes demonstrate, the richness of cultural skills, diversity and care for nature is deeply embedded in the collective practices of the locals. As more responsible agency emerges through relationships, meaningful engagement among diverse stakeholders becomes a key determinant for the community to thrive. The co-creation of heterotopian communities therefore necessitates a collaborative approach where the voices, practices and aspirations of the locals are central to shaping the local identity, practices and narratives. Advancing the knowledge on the recognition of shared local values and traditions as vital for community agency (Guri *et al.*, 2021), we learn from Ana, a founder of the Association of citizens 'Milanovac Ethno Wellness', who explained:

Our Group is linked to ethnology and anthropology, and I wanted to make that connection more apparent. However, merely gathering women engaged in various handcrafted activities, from wool work to making jams, wasn't sufficient for me. Because we also have the Old Oak, we were like let's tell the story about it. I have a deep appreciation for traditions, especially those intertwined with nature. This relates to local toponymy, for example, Treska Mountain, my villages are called Jablanica and Grabovica [*after trees*]. The home to an ancient oak, possibly the oldest in Serbia, is situated in my village. Everything somehow revolves around traditions and nature. It might sound like a fairy tale, but there's a very deep meaning in it all.

At the core of the community agency is the cultivation of their capacity to access and utilise resources at hand, while facilitating their joint commitment to community development. Ana's narrative portrays a community as a space that is represented by diverse stakeholders (including nature), embodying creative practices through various handcrafted activities. However, the mention of the Old Oak adds a layer to the narrative, suggesting a connection to the deep-rooted history and natural environment, echoing the principles of new materialist thinking that emphasise the significance of relations with non-human entities (Kortetmäki et al., 2022). The attachment to local toponymy and the reference to mountains and villages further emphasise the relational uniqueness of histories, traditions and nature. Ana's description reflects a heterotopian space where the community's activities and narratives encourage both residents' and visitors' entanglements with Šumadija's traditions and the natural resources with which the locals live and learn from, as illustrated in the quote:

> We notice changes, we know what kind of trees live in the area, I know medicinal plants, I do not know mushrooms, which may have remained my wish to understand better. We carry hot water in a thermos and make tea in the meadow. This way we're always providing extra enjoyment during the outing.

Ana highlighted making tea in a meadow as a creative activity intertwined with learning about the land, traditions of local communities and their local knowledge. The observation of changes in the environment and her awareness of the types of trees in the area reflect a deep connection to the land. The act of carrying hot water in a thermos and making tea in the meadow becomes a ritual that enhances the experience of those who engage in ethno-wellness practices. This practice is slow and cultural, where everyday activities, such as making tea, might become meaningful cultural tourism practices that contribute to a holistic understanding of the local landscapes, plants and traditions. Kosta revealed that he is also engaged in creative activities through woodwork deeply rooted in nature, which represents an additional income:

> I also do some wood craft and leather craft. I make spoons and cups from wood. I saw my own leather sheets for axes and knives and other stuff. So I make everything from nature, take from nature and give back to the nature, live with the nature like Native Americans did, you know, everything today is too modern, too easy, too fast.

The act of making objects from natural materials echoes a desire to live in harmony with nature, reminiscent of traditional practices observed in

Native American cultures. Kosta's critique of modernity, characterised by its speed and convenience, tells of his commitment to a slower, more deliberate way of life that aligns with the principles of creative slow cultural tourism. Through these creative activities, and sharing them with others (locals or participants on tours), he contributes to the preservation of traditional craftsmanship, fostering a deeper connection between humans and the cultural and natural heritage of the region. Another illustrative example of creative collaboration with local stakeholders is evident in Filip's aspirations to open a restaurant within the camp. Instead of pursuing this idea independently, he turned to collaboration with his neighbours, who mastered the craft of preparing homemade food for the camp visitors:

> I thought of opening my own restaurant, so that people could enjoy homemade products, even if they weren't staying at the camp. However, I decided to start by collaborating with my neighbours to provide breakfast. These long-time villagers could use some financial support, and I believe it's a win-win situation – satisfying both the locals and the guests. After all, who could resist the charm of homemade pie or a warm bun delivered to their doorstep at 9 am?

Significant attention is paid to stakeholder collaboration and community engagement in the development of tourism initiatives. Initially contemplating the idea of opening his own restaurant, Filip recognised the potential for providing homemade products to a broader audience beyond the camp visitors. However, opting for partnering with neighbours to offer breakfast reflects a thoughtful consideration of the community's needs and a recognition of the financial support that long-time villagers could benefit from. Such collaboration with locals, Filip not only supports the economic well-being of the community but also creates benefits for all. This approach exemplifies the importance of involving and empowering local stakeholders, aligning slow cultural tourism with the principles of sustainable development in that it prioritises community well-being, economic development and the preservation of local cultures through homemade products, such as pies or warm buns, thus enacting sustainable practices and cultivating ethics of care (Sama *et al.*, 2004).

Slowness

Slow cultural tourism alludes to the slow exploration of cultural elements, emphasising depth, meaningful engagement in and with the activity, and a connection with the local environment. The heterotopian community is very much connected to its unique temporal and spatial characteristics. For example, the idle pace of exploration and the emphasis on qualitative, kairological time that Farkić *et al.* (2023) referred to as a crucial dimension of wellbeing, might contribute to our understanding of the dynamic and authenticity of emergent relations within the community. Reflecting on some idle moments entwined with the creative act of carving wood, Marko reminisced:

> I carved a stick, you know… a twig, and I sat on the top of a hill for three hours doing just that. Some would say he's wasting his time, he doesn't know what he's doing, but I did it for three hours and had a fantastic time. You let it lead you… I think about thoughts, I think about where my thoughts come from and where they will take me. Those patterns you create on the piece of wood are just to see where you are from the inside, some kind of introspection… Some people think that shepherds are stupid people, but they are not, they are wise people. Being alone in the same place for days, decades, without succumbing to boredom or insanity reflects remarkable inner stability.

Slowness, as depicted in the quote, intertwines with cultural values that celebrate contemplation, creativity, connection to nature and the wisdom gained through solitude and solipsistic moments. It challenges external judgments and emphasises the intrinsic worth of activities that may be considered slow-paced but hold deep cultural significance in bolstering inner stability and personal fulfilment. The act of grounding through carving a stick for three hours contrasts with a fast-paced, goal-oriented mindset often associated with modern lifestyles. An emphasis here is placed on idle manual work, creativity and the intrinsic value of creating something with one's hands, the process that can be valorised in slow cultural tourism. Kosta's insights also contribute to our understanding of how *natureculture* intertwines with slowness. The act of working with technologies and material, such as tools and wood, is described as both relaxing and challenging, and requires patience and careful attention, while the process itself and the depth of experience take precedence over speed and efficiency:

> When you're out in nature, you have to slow down. Like last week, we were making the shelter together. A friend suggested to bring an electric hand drill. And I told him, no, we don't need that, we will do it manually. [...*explains the process of using a manual drill*...] And yeah, sure everything is slower. But when you do it, you feel better about it. You know, you enjoy the process. Yeah, you've put some hard work into it. Especially working with wood. It's very relaxing. It can be stressful when you make a mistake, because you can't put the wood back if you break it it's broken, you need another piece… But it's something you take your time with… We were making bushcraft shelter… the whole day, you get up in the morning, you go you chop down the wood, you cut it to size, you put up a construction, you use a hammer, you use an axe, you knock down the nails…when we are out you are not looking at your watch all the time, I'm taking it slow I'm not rushing anything…we were just taking it slow and enjoying it…

Relational Futures and Slow Cultural Practices

In this account, I aimed to join the scholarly efforts to unsettle the abstract narratives of the Anthropocene by maintaining focus on the possibilities of engaging differently with ordinary, everyday and multiple relations within the context of slow cultural tourism (Rantala *et al.*, 2020). In the time of Earthly crises, cultural tourism needs to play its role in facilitating a deeper understanding of the more-than-human knowledge and values. It has the potential to bridge the gap between human communities and the natural world, cultivating an appreciation for the ecological wisdom embedded in diverse cultures globally. This study has focused on a community living in the forested heart of Serbia, members of which made a deliberate choice to displace themselves from what Max Weber calls the iron cage of modernity. Engaged in tourism development in diverse ways, they enact sustainable practices and utilise traditional ecological knowledge, while maintaining symbiotic relationships with their environment. Such relational engagement is reflected in tourism products that they co-produce and instil values of environmental stewardship, cultural understanding and ethical way of living and being.

My aim in this chapter was to illustrate how we can learn with and through the complex processes which are constantly emerging and taking place in and with natural environments (physically) and cultural tourism (conceptually), through touristic experiences. As Rantala *et al.* (2020) proposed, we should sensitise ourselves to the new stories that are being born every moment and to the new histories that are being created. Attuning to the stories of one community, I was able to understand the way in which slow cultural tourism might entangle matter, materials, artefacts in consumption and production, histories, local communities, narratives or tourists. To make sense of these relations, I built upon the post-anthropocentric ideas in taking the feminist new materialist approach, while maintaining my focus on the emergence of so called heterotopian communities. The knowledge produced here, however, is situated, partial and always in becoming.

A community-based approach to tourism development in natural areas offers a promising framework for cultural tourism. To envision its future, especially within the context of community-driven slow cultural tourism, it is essential to explore the dynamics and emergent phenomena within pockets of resistance. The valuable insights derived from conversations with four community members actively involved in tourism development in Šumadija enable us to understand the Anthropocene's impact and may stimulate contemplation on the choices we make in an increasingly volatile, uncertain, complex and ambiguous environment. It is noteworthy that the empirical insights captured through this study span anthropological, archaeological, ethnological, economic, physical education and tourism perspectives. However, they are all rooted in the shared aspiration to live life ethically, meaningfully and authentically in the emergent heterotopias. It is therefore hoped that heterotopian communities and bubbles of ethical consumption and (knowledge) production, will materialise across different spaces and places, whether in the form of tourism or culture cooperatives or slow scholarship communities, be it at the Serbian hillsides of Ležimir or Irig, Saint-Raphaël, or elsewhere on the Planet Earth.

References

Barad, K. (2007) *Meeting the Universe Halfway: Quantum Physics and the Entanglement of Matter and Meaning.* Duke University Press.

Beames, S., Mackenzie, S.H. and Raymond, E. (2022) How can we adventure sustainably? A systematized review of sustainability guidance for adventure tourism operators. *Journal of Hospitality and Tourism Management* 50, 223–231.

Boukhris, L. (2020) Decolonizing natural heritage: Knowledge, power and the political economy of tourism. In M. Gravari-Barbas (ed.) *A Research Agenda for Heritage Tourism* (167–182). Edward Elgar Publishing.

Braidotti, R. (2013) *The Posthuman.* Polity Press.

Bui, H.T., Jones, T.K., Weaver, D. and Le, A.V. (2020) The adaptive resilience of living cultural heritage in a tourism destination. *Journal of Sustainable Tourism* 28 (7), 1022–1040.

Caffyn, A. (2012) Advocating and implementing slow tourism. *Tourism Recreation Research* 37 (1), 77–80.

Calvi, L., Moretti, S., Koens, K. and Klijs, J. (2020) The future of cultural tourism: Steps towards resilience and future scenarios. *ENCATC* 2, 34–40.

Farkić, J., Isailović, G. and Lesjak, M. (2023) Conceptualising tourist idleness and creating places of otium in nature-based tourism. *Academica Turistica-Tourism and Innovation Journal* 15 (1), 11–23.

Farkić, J., Filep, S. and Taylor, S. (2020) Shaping tourists' wellbeing through guided slow adventures. *Journal of Sustainable Tourism* 28 (12), 2064–2080.

Foucault, M. (1986) Of other spaces. *Diacritics* 16 (1), 22–27.

Fox, N.J. and Alldred, P. (2016) *Sociology and the new materialism.* Sage.

Fullagar, S., Markwell, K. and Wilson, E. (eds) (2012) *Slow Tourism: Experiences and Mobilities.* Channel View Publications.

Greenwood, D.A. (2013) Culture, environment, and education in the Anthropocene. In M.P. Mueller, D.J. Tippins and A.J. Stewart (eds) *Assessing Schools for Generation R (Responsibility) A Guide for Legislation and School Policy in Science Education* (pp. 279–292). Springer Netherlands.

Grimwood, B.S., Stinson, M.J. and King, L.J. (2019) A decolonizing settler story. *Annals of Tourism Research* 79, Article 102763.

Guri, E.A., Osumanu, I.K. and Bonye, S.Z. (2021) Eco-cultural tourism development in Ghana: Potentials and expected benefits in the Lawra Municipality. *Journal of Tourism and Cultural Change* 19 (4), 458–476.

Hall, C.M., Baird, T., James, M. and Ram, Y. (2016) Climate change and cultural heritage: Conservation and heritage tourism in the Anthropocene. *Journal of Heritage Tourism* 11 (1), 10–24.

Haraway, D. (2013) *Simians, Cyborgs, and Women: The Reinvention of Nature.* Routledge.

Harrison, R. (2015) Beyond 'natural' and 'cultural' heritage: Toward an ontological politics of heritage in the age of Anthropocene. *Heritage & Society* 8 (1), 24–42.

Hartley, D. (2016) Anthropocene, Capitalocene, and the problem of culture. In J.W. Moore (ed.) *Anthropocene or Capitalocene? Nature, History, and the Crisis of Capitalism* (pp. 154–165). PM Press.

Higgins-Desbiolles, F., Carnicelli, S., Krolikowski, C., Wijesinghe, G. and Boluk, B. (2019) Degrowing tourism: Rethinking tourism. *Journal of Sustainable Tourism* 27 (12), 1926–1944.

Höckert, E., Kinnunen, V. and Rantala, O. (2023) Suggestions for future wanders. In O. Rantala, V. Kinnunen and E. Höckert (eds) *Researching with Proximity: Relational Methodologies for the Anthropocene*. Palgrave Macmillan.

Holden, A., Jamal, T. and Burini, F. (2022) The future of tourism in the Anthropocene. *Annual Review of Environment and Resources* 47, 423–447.

Kortetmäki, T., Heikkinen, A. and Jokinen, A. (2023) Particularizing nonhuman nature in stakeholder theory: The recognition approach. *Journal of Business Ethics* 185 (1), 17–31.

Lather, P. (2016) Top ten+ list: (Re)thinking ontology in (post) qualitative research. *Cultural Studies – Critical Methodologies* 16 (2), 125–131.

Mandarić, M., Milićević, S. and Sekulić, D. (2017) Traditional values in the function of promotion of Šumadija and Pomoravlje as rural tourism destinations. *Economics of Agriculture* 64 (2), 787–803.

Matarrita-Cascante, D., Brennan, M.A. and Luloff, A.E. (2010) Community agency and sustainable tourism development: The case of La Fortuna, Costa Rica. *Journal of Sustainable Tourism* 18 (6), 735–756.

Matteucci, X., Koens, K., Calvi, L. and Moretti, S. (2022a) Envisioning the futures of cultural tourism. *Futures* 142, Article 103013.

Matteucci, X., Nawijn, J. and von Zumbusch, J. (2022b) A new materialist governance paradigm for tourism destinations. *Journal of Sustainable Tourism* 30 (1), 169–184.

Monaheng, T. (2016) Integrating indigenous knowledge in the development of cultural tourism in Lesotho. In H. Manwa, N. Moswete and J. Saarinen (eds) *Cultural Tourism in Southern Africa* (pp. 31–46). Channel View Publications.

Mzembe, A.N., Koens, K. and Calvi, L. (2023) The institutional antecedents of sustainable development in cultural heritage tourism. *Sustainable Development* 31 (4), 2196–2211.

Oh, H., Assaf, A.G. and Baloglu, S. (2016) Motivations and goals of slow tourism. *Journal of Travel Research* 55 (2), 205–219.

Prasetyo, N., Filep, S. and Carr, A. (2023) Towards culturally sustainable scuba diving tourism: An integration of Indigenous knowledge. *Tourism Recreation Research* 48 (3), 319–332.

Rantala, O., Salmela, T., Valtonen, A. and Höckert, E. (2020) Envisioning tourism and proximity after the Anthropocene. *Sustainability* 12, Article 3948.

Saari, R., Höckert, E., Kugapi, O., Lüthje, M. and Mazzullo, N. (2020) Cultural sensitivity in Sámi tourism: A systematic literature review in the Finnish context. *Matkailututkimus/Finnish Journal of Tourism Research* 16 (1), 93–108.

Sama, L.M., Welcomer, S.A. and Gerde, V.W. (2004) Who speaks for the trees? Invoking an ethic of care to give voice to the silent stakeholder. In S. Sharma and M. Starik (eds) *Stakeholders, the Environment and Society* (pp. 140–165). Edward Elgar Publishing.

Su, M. and Wall, G. (2015) Community involvement at Great Wall World Heritage sites, Beijing, China. *Current Issues in Tourism* 18 (2), 137–157.

St. Pierre, E.A. (2019) Post qualitative inquiry in an ontology of immanence. *Qualitative Inquiry* 25 (1), 3–16.

Varley, P. and Semple, T. (2015) Nordic slow adventure: Explorations in time and nature. *Scandinavian Journal of Hospitality and Tourism* 15 (1–2), 73–90.

Viken, A., Höckert, E. and Grimwood, B.S. (2021) Cultural sensitivity: Engaging difference in tourism. *Annals of Tourism Research* 89, Article 103223.

Wengel, Y. (2021) The micro-trends of emerging adventure tourism activities in Nepal. *Journal of Tourism Futures* 7 (2), 209–215.

7 Emerging Perspectives on the Future of Cultural Tourism

Dallen J. Timothy

Introduction

For more than a century, since the Industrial Revolution, tourism has grown commensurate with the increased affluence of Western societies. As societies became more prosperous and modernised, people's desires to discover the world also grew and international travel has come within reach of increasing numbers of people. In 2012, for the first time in history, international tourist arrivals reached the 1 billion mark. Since that time, with the exception of the jolt of the COVID-19 pandemic in 2020–2022, tourism has grown steadily. The year 2023 saw approximately 1.3 billion international trips, and forecasts suggest that 2024 will see the full recuperation of tourism to pre-pandemic levels (UN Tourism, 2024).

A significant portion of global travel is comprised of consuming heritage and cultural experiences. Cultural tourism and heritage tourism are estimated to comprise somewhere between 40–70 percent of all international travel (Timothy, 2021a; UNWTO, 2018). Some of the most iconic symbols of tourism include heritage sites of global acclaim, such as Angkor Wat, the Great Wall of China, Machu Picchu, Borobudur and the Roman Forum, to name only a few. These renowned locales are made increasingly famous as their images are used by national tourism organisations and are branded UNESCO World Heritage Sites, which does not automatically increase tourism (Hall & Piggin, 2002; Iațu et al., 2018), but some studies show that the UNESCO brand does have the potential to increase tourism at certain sites (Adie & Hall, 2017). Nonetheless, the WHS designation is frequently capitalised on as a brand and marketing tool by countries that wish to grow their heritage tourism sector (Adie, 2017).

With the overwhelming focus on cultural heritage by mass tourism, historic cities and other heritage places have born the brunt of tourism's negative outcomes. This has largely been the way of cultural tourism, but several new trends are emerging that have the potential to challenge these traditions of mass tourism. This chapter will examine several trends in current and future cultural tourism practice within the framework of sophistication of travel demand, deeper experiences, and cultural globalisation.

Dissatisfaction with Mass Tourism

Since the Second World War, mass tourism has grown exponentially with increasing footfall in the most popular destinations, causing discontent among residents and tourists and exacting significant sociocultural and ecological impacts that are hard to overcome (Hernandez-Maskivker et al., 2021). Overtourism is now the norm in many tourist destinations, including some of the most iconic heritage places, such as Rome, Venice, Barcelona, Paris, Prague, Beijing, Bangkok and Kyoto (Dodds & Butler, 2019), although several of these cities have enacted policies and practices to lessen mass tourism's local impacts, including limiting visitor numbers and charging fees to enter the city (e.g. Venice). Cultural heritage-based tourism accounts for much of this trend in overtourism, resulting in deep negative outcomes not only for destination residents but also for the built environments and intangible cultures consumed by visitors.

In response to the diminished experience of mass tourism for both residents and tourists, as well as increased global mobility, changes in technology and growing alternative tourisms, special-interest travel has grown parallel to mass tourism, sometimes replacing it for many travel consumers. This has resulted in the immense growth in unique tourism niche markets that seek experiences beyond the normative mass tourist encounter (Cetin & Bilgihan, 2016; Garau Taberner, 2009; Jovicic, 2016). This manifests in specific types and subtypes of tourism that are defined both by travellers' desired outcomes and the types of activities they undertake or sites they visit in the destination. Prominent examples include volunteer tourism, ecotourism and heritage tourism, with heritage tourism encompassing many other subtypes such as dark tourism, religious and spiritual tourism, agritourism and literary tourism, to name only a few.

These changes, together with increased affluence, higher levels of education over previous generations, increasingly adventurous personalities, a growing awareness of the world through personal travel experiences and those of others posted on YouTube and other social media platforms, an increase in general environmental consciousness, and a growing sense of altruism, have led to a sophistication of tourism demand (Seeler et al., 2021). This is especially notable in the realm of cultural tourism (Gheorghilaş et al., 2017; Timothy, 2018), where many travellers are now less satisfied with visiting only the conventional, iconic cultural experiences and heritage places. Instead, they seek experiences that expose them to the everyday life of ordinary people in different cultural settings (Timothy, 2014a, 2014b). Mass tourism experiences are increasingly seen as superficial and stereotypical (Jovicic, 2016; Harrison & Sharpley, 2017) in a world that is becoming hungrier for deeper existential experiences in travel (Chambers, 2009; Santos *et al*., 2016).

Although most travellers still desire to visit globally-recognised famous heritage places, the sophistication of demand is leading to different expectations in experience and product design. Many tour companies and destination management organisations have realised the lucrativeness of alternative markets and experiences, and have started to plan and provide atypical opportunities for travel consumers who have higher expectations of participating in something unique, extraordinary, equitable and sustainable. These modern patterns and trends reflect many of the utopian characteristics outlined by Matteucci and his colleagues (2022), including degrowth, slower and more meaningful experiences, authentic encounters, collective wellbeing and increased social and environmental justice. Because there will be no single future as such (Urry, 2016), but a combination of cultural tourism futures, Matteucci *et al*. (2022: 6) suggest that the future will be heterotopian in that 'pockets of resistance' to mass cultural tourism will multiply within a slowly decaying neoliberal order. They describe 'pockets of resistance' as alternative spaces where more ethical (utopian) practices thrive. The following sections highlight some of the outcomes and trends related to this paradigmatic shift from dystopian tourism practices to a greater emphasis on sustainability, wellbeing, existential transformation and democratised processes.

Deeper Experiences

As previously noted, a significant part of the change in tourists' attitudes denotes a breakaway from the often-superficial experiences of mass tourism towards a greater desire for deep and more authentic experiences – authenticity itself being a hotly contested topic but a sentiment that is usually determined by each individual based on their own needs, interests and backgrounds (Timothy, 2021a; Wang, 1999). There are many ways in which this sense of existential authenticity manifests in people's travel desires, including slow and immersive tourism, a greater appreciation for ordinary heritage, transformational tourism, co-creative tourism and vicarious travel. Each of these is examined briefly below.

Slow and immersive tourism

The slow food movement grew in response to the unhealthy, harried lifestyles associated with eating quickly and consuming large quantities of fast food. It encourages people to eat healthier, become familiar with the sources of their food, appreciate culinary heritage, and slow down in their gustatory practices (Boyd, 2016). Slow tourism emerged from the slow food and Cittaslow movements as a reaction to the fast-paced lifestyles of most tourists who try to fit inordinate amounts of activities and places into an itinerary, which enables them only to scratch the surface of a destination, experiencing places and people in only a superficial sort of way. Slow tourism, on the other hand, entails people spending more time in the destination, slowing down, delving deeper into local cultures, becoming more educated about the destination and taking the time to appreciate their natural surrounds in greater measure. Thus, slow tourism emphasises quality over quantity and deeper, more meaningful experiences over the fleeting and shallow activities that are most closely associated with the 'typical tourist'.

In response to this trend, destinations and entrepreneurs have created opportunities to 'live like a local' (Mkhize & Ivanovic, 2020), even if some of these opportunities are rather staged or invasive. Likewise, not all elements of immersive tourism can be viewed as positive. For example, poverty immersion may be a mix of positive and negative. While spending a night and dining with a family in a Brazilian favela or a South African township may create memorable experiences for the tourist and provide sustenance for the host family (Mkhize & Ivanovic, 2020), some observers see this as the 'same old tourism' that commodifies people's misfortunes and accrues wealth for the tour companies and other mediators rather than the families who live in the ramshackle squalor that is sold to tourists (Altamirano, 2023; Frisch, 2016). By gratifying the voyeuristic desires of tourists, we bear the risks of engaging in a dystopian 'aestheticization of poverty, producing a depoliticised image that becomes attractive for capitalist exploitation' (Dovey & King, 2012: 290).

Despite the negative 'mass-touristification' of certain types of slow and immersive tourism, there is growing demand for more unique and immersive experiences that enable visitors to have deeper, more eudaemonic experiences that engage more holistically with the destinations they visit and which bring mutual benefits for the communities that host them (Seeler *et al.*, 2022). For the most part, these have positive outcomes, build bridges and create greater cross-cultural understandings.

A greater appreciation of ordinary heritage

A manifestation of slow and immersive tourism is evidence of an increasing interest in ordinary heritage over (or in addition to) the grandiose heritage that tourism has focused on in the past (Podder *et al.*, 2018; Timothy, 2020, 2021b). For a long time, tourism has focused overwhelmingly on the heritage of the extraordinary: the kings and queens, military leaders and national heroes, scientists, poets and artists, explorers and colonisers, and famous musicians, largely due to the Eurocentric approaches to heritage valuation and management in colonial societies. Certain manifestations of rural heritage and folk museums have been the exception. Yet, in the greater scheme of things, the extraordinary heritage of these elites is minuscule compared to the heritage of the billions of inhabitants who have heretofore dwelt on this planet. The democratisation of heritage is becoming more regular in more places. The growing emphasis on the slaves rather than only on the slave owners in the American South, is one prominent example in recent decades.

People's growing interest in knowing how ordinary people lived throughout history, rather than seeing heritage as only the purview of the rich and famous, has created greater demand for ordinary heritage, such as agriheritage, rural landscapes, religious sites, living cultures, Indigenous practices, culinary traditions, linguistic practices, vernacular architecture and folk museums. No longer are the world's greatest monuments, many of which are UNESCO World Heritage Sites, standalone attractions for all travel segments. Increasingly, knowledgeable travellers are keen to appreciate the living cultures of the sites' descendent communities who continue to live in their vicinity and whose literal heritage is on display.

Transformative tourism

Transformative tourism refers to changes in people's lives through impactful, meaningful and evocative experiences that affect lifestyle, behavioural, or belief system changes in areas such as spirituality, learning, self-confidence, wellbeing, or any number of other potential areas of personal life (Chhabra, 2021; Kirillova et al., 2017; Sheldon, 2020). A spiritual awakening through communion with a deity in sacred space or experiencing the wonderment and awe of nature through deep outdoor experiences may be exemplary transformative tourism events. According to Pung et al. (2020), transformative encounters provoke peak episodes and dilemmas, perhaps through culture shock, challenging situations in the destination, interacting with local residents, especially those in less-affluent parts of the Global South, or complete awe and utter amazement at the beauty of a landscape visited. Transformative tourism pushes tourists beyond their comfort zones, educes more inclusive worldviews, and promotes cross-cultural understanding (Soulard et al., 2021). Its main premise is that the transformations are long-lasting, beyond the duration of the journey (Pung & del Chiappa, 2020). These context-specific stimuli may lead people to reflect, interpret, change their perspectives and values, or alter their habits and behaviours temporarily or permanently. For many travellers, these experiences are life-changing (Matteucci, 2022; Teoh et al., 2021).

The most often cited forms of tourism with transformative power include volunteer tourism, educational tourism, religious tourism, wellness tourism, nature-based tourism and cultural tourism (Coghlan & Weiler, 2018; Teoh et al., 2021; Zhao & Agyeiwaah, 2023) – all providing once in a lifetime experiences of self-discovery or altruism. Cultural/heritage tourism plays a crucial role in transformative tourism. For example, visiting a Holocaust or genocide museum can have such an impact on visitors, that their beliefs, behaviours and everyday action spaces maybe challenged and modified to be more sensitive and empathetic (Vidickiené *et al.*, 2020). Likewise, immersive experiences in rural and less-affluent areas can be hugely transformative (Vidickiené *et al.*, 2020), although short stays, contentious travel companions, repetitious activities, and lack of access to local lifestyles and cultures can act as inhibitors to transformative experiences (Pung & del Chiappa, 2020).

In the realm of religious tourism (or pilgrimage), which is a pure manifestation of heritage tourism, encounters of faith, facing personal crises, discovering one's purpose in life, communion with a higher power, undertaking sacred rituals and humbling oneself can have lasting effects (Collins-Kreiner, 2020; Liutikas, 2021). The transformative process of the Camino de Santiago pilgrimage trail has been documented equally among Roman Catholics and non-religious users who hike the trail as an exercise in self-discovery, cultural appreciation or simply personal reflection (Barlar, 2021). Likewise, spiritual retreats and related practices, such as yoga, silence and meditation provide holistic experiences for the whole self (Schedneck, 2021). The personal quest in spiritual and religious tourism is often spoken of in transformative terms through authentic, self-aware and immersive experiences (Mkhize & Ivanovic, 2020).

Personal heritage-based tourism also often educes significant individual transformations. For many people, roots-seeking becomes a journey of complete self-actualisation. By visiting the lands of their ancestors, walking where their forebears walked or seeing where they suffered, and undertaking family history research in situ, people often discover who they are, where they come from, and why they are here (Timothy, 1997, 2008). Many people describe it as a spiritual experience and one that can prove cathartic or, on the other end of the spectrum, spark feelings of bitterness, anger and revenge (Timothy & Teye, 2004).

Co-creating heritage tourism

Another related trend affecting the future of cultural tourism is co-creative tourism. Co-creative tourism (or creative tourism) refers to destination residents/service providers and tourists working together actively to create heritage values and meaningful experiences. It is an interactive process in which service providers, residents and tourists co-design touristic encounters through active learning and the exchange of information (Cabeça *et al.*, 2020). According to Carvalho *et al.* (2023), co-creation emphasises tourists' active role in generating value from interacting with people and places in the destination which, according to Sugathan and Rakesh Ranjan (2019), can affect a tourist's experience and influence their decisions whether or not to return. In the realm of cultural heritage, co-creative principles include engagement, interaction, participation and personalisation to create meaningful and immersive experiences. Matteucci and Smith (2024: 93) have recently presented the creative tourist experience as a 'manifestation of empowerment' and as 'a space of resistance with emancipatory powers'. From a future perspective, Matteucci *et al.* (2022) have positioned creative tourism activities as alternative, heterotopian spaces of cultural heritage consumption.

Even in the context of archaeology-based heritage tourism, co-creative strategies can be implemented to promote participative heritage interpretation that values tourists' input and understanding of the past to help create deeper meanings and heritage values (Ross, 2020). These sorts of co-created experiences take tourists' visits beyond the mere tourist gaze, to stimulate greater levels of curiosity and engage with destination cultural heritage in less sterile ways than simply visiting museums and heritage centres (Kastenholz & Gronau, 2022). This manifests well in the idea of public archaeology in which the public is more highly invested in cultural site preservation and use. Likewise, situating the self in cultural contexts and heritage experiences can be a significant way of breathing new life into cultural heritage visits, interpretive programs and interactive communication (Breathnach, 2006; Poria *et al.*, 2006). According to Weiler and Black (2022), the increasingly sophisticated tourism marketplace is now beginning to expect their guides and interpreters to be collaborators in the heritage experience rather than simply communicators.

Vicarious tourism

Although proxy travelling through the experiences of others, involving story sharing, slideshows, looking at souvenirs and trip mementos and watching television shows have long been an important part of surrogate travel and travel sharing (Chandralal *et al.*, 2015; Marder *et al.*, 2019; Reiss & Wiltz, 2004), a current trend is vicarious travel through social media influencers. In recent years there has been a surge of social media output that focuses on travel and cultural heritage consumption in particular (Lin & Rasoolimanesh, 2023; Price & Kerr, 2018). For example, there are now thousands of YouTube channels that focus on globetrotting, with the growing popularity of vloggers providing content that was essentially unavailable only a decade ago. One recent trend on YouTube and other social media is visiting as many countries as possible, or visiting every country in the world. Several people have achieved the aim of visiting every country on the planet, garnering many views and likes. Others focus on visiting the most dangerous places on Earth, including war zones, or last-chance destinations, such as the 'sinking' islands of the Pacific and arctic environments that are on the verge of disappearing. People can live vicariously through the experiences of YouTubers and other social media content creators who travel the world and share their cultural encounters with their followers.

One such channel the author regularly watches is 'Itchy Boots', which features a Dutch woman (Noraly) travelling the world alone on her motorcycle. She lives full-time on her motorbike and in local guesthouses. She spends most of her time avoiding paved highways, instead taking backroads and deeply immersing herself in the living cultures of the places she visits. At the end of 2023, she had vlogged in more than 65 countries in Asia, Africa, Europe, North America and Latin America. In 2021–2022, she completed a journey from the southern tip of South America to the north coast of Alaska. At the time of writing, Noraly is currently biking the length of Africa along the continent's west coast. Comments on the Itchy Boots channel clearly demonstrate vicarious globetrotting: 'Noraly takes me places I never would have dreamed of going before'; 'I live through Noraly's experiences'; 'I have mobility problems and could never attempt go to the places she goes'; and 'I can't wait to see what countries we will all be going to next season with Itchy Boots'. These sentiments reflect not only people's admiration of Noraly's fortitude and sense of adventure but for the vicarious way in which they too can experience far-flung parts of the planet they might otherwise not have a chance to visit. These sorts of YouTube channels focus largely on immersive cultural experiences. Even those whose main purpose is to collect the most countries devote a significant amount of time to tangible and intangible cultural experiences. The expenses incurred and the risks taken by explorers such as Noraly take the costs and risks associated with travel from 'armchair travellers' but enrich their lives through the proxy experiences of online influencers.

Cultural Globalisation

Most treatises on globalisation and its various processes focus on economic and geopolitical domains, yet there is also cultural globalisation (Hopper, 2007). Music, television shows, art, movies, video games and social media are examples of cultural globalisation, wherein elements of culture (including pop-culture) are disseminated and consumed worldwide. Pop culture-based tourism is one example of cultural globalisation (Yamamura & Seaton, 2020). The increasing popularity of Chinese music and dance across the Western world in the past decade and Irish dance touring globally, as well as the widespread popularity of Korean and Mexican television dramas and soap operas, are examples of this increasingly widespread phenomenon. With increased global connectedness through music, television, movies and social media, fans are consuming destinations and events associated with their favourite pop culture icons and pop culture forms (Olsen, 2021; Timothy, 2019). The popularity of American baseball in Japan and the increasing availability of diverse ethnic foods in various places are also now-commonplace indicators of cultural diffusion brought about by many forces, including tourism

(Timothy, 2019). The growing fanbase of Afrobeats music outside the African diaspora (Krings & Simmert, 2020), the afición or passion for flamenco music and dance outside Spain (Matteucci, 2013), the popularisation of Buddhist temple food among vegetarians, vegans and healthy eaters worldwide (Son & Xu, 2013), and the greater appreciation of animist beliefs and spiritual traditions in Africa among non-religious visitors (Gedecho & Nyikana, 2023) all attest to the growing influence of cultural globalisation.

Cultural globalisation has enabled people to thrive better in different environments, so that they are more comfortable in culturally diverse situations (Chambers, 2009). Likewise, cultural globalisation has made very local places and cultural practices global through social media, television, and other means of globalisation. Thus, in these and other respects, cultural globalisation has accelerated people's mobility along Plog's (1974) scale of psychocentricity and allocentricity from being more inward-looking and fearful of new cultural settings (psychocentric) to being more accepting of cultural differences and a deeper desire to undertake journeys to places of Otherness. While globalisation has led to a profusion of cultural identities and to the hybridisation of different cultural forms (Smith, 2016), it has also been blamed for standardising and homogenising cultures. In the long run, very few places and cultural expressions will remain untouched by globalisation (Ritzer, 2003), and critics of globalisation anticipate a rather dystopian future in which a cultural trip will take consumers to the shopping mall.

Conclusion

Cultural tourism, especially heritage tourism as described in this chapter, has long been a mainstay of mass tourism since the Industrial Revolution. The tangible heritage and living cultures of other places have appealed to travelling consumers for centuries and are the focus of much tourism today. The dystopian massification of cultural heritage-based tourism has resulted in degraded environments and unsatisfactory individual experiences (Matteucci et al., 2022). Likewise, mass tourism has largely become antithetical to the principles of sustainable tourism development. Thus, with changing public discourses on sustainability, degrowth, environmental and social justice, healthier living, and other such issues of contemporary times, we have seen the beginnings of a significant shift from pure mass consumption to more sophisticated demand and selective experience-seeking. Deeper, slower and more immersive experiences derive from this move away from mass consumption on an individual level. This shift may lead the way towards more utopian or heterotopian futures. Seeing how ordinary people live, not only the rich and famous, has now become vogue. Seeking transformative experiences, co-creating personal cultural encounters and living vicariously through the adventures of others are now familiar practices made possible largely through the processes of cultural globalisation. These processes and changes are becoming increasingly notable in nearly all facets of tourism and will continue to transform the industry as destinations seek to diversify their products and as diverse markets demand increasingly transformative experiences and other encounters that challenge the traditions of tourism that have dominated the sector until now.

These contemporary changes and manifestations of dissatisfaction with traditional tourism have manifested in both altruistic and self-absorptive motives underlying travel decisions. In the former category, we can see the emergence and rapid growth in volunteer tourism, education travel, solidarity tourism and several types of pilgrimages. These experiences are often underscored by principles such as adaptability, resilience, transformative experiences, existential journeys, sensitivity to the plight of others, the betterment of the planet and positioning the self in the context of the greater good of society. As such, these contemporary changes are socially valuable, and it is hoped that utopian cultural tourism futures will unfold. The tendency toward self-absorption often entails seeking peak experiences, self-actualisation and discovery, efforts to outdo others, and undertaking fandom pilgrimages that reflect the exponential growth in pop culture through social media and other globalising forces as people travel to see the places associated with games, soap operas, and music genres, or to meet their favorite influencers (Alderman, 2002; Beeton, 2019; Digance, 2006; Olsen, 2021). With more widespread engagement with content creators and the growing influences of cultural globalisation, as well as the growing need for altruistic encounters with the Other, consumers will continue to seek travel experiences that either indulge or contribute to the greater good of society.

The vicissitudes described in this chapter point to an overall current and future re-assessment of the values ascribed to cultural tourism. Personal connectedness, self-actualisation and a realisation of one's own place in the world underscore many of the positive changes we now see in the individualisation of cultural tourism and it is certain to continue as people continue to seek their rootedness in an increasingly tumultuous world. There is much scope in future research to understand these changes in far greater depth. How destinations will cope with changing demands for cultural heritage experiences is essential in moving forward on the path to more sustainable, immersive, creative, transformative and disruptive cultural tourism experiences.

References

Adie, B.A. (2017) Franchising our heritage: The UNESCO World Heritage brand. *Tourism Management Perspectives* 24, 48–53.

Adie, B.A. and Hall, C.M. (2017) Who visits World Heritage? A comparative analysis of three cultural sites. *Journal of Heritage Tourism* 12 (1), 67–80.

Alderman, D.H. (2002) Writing on the Graceland wall: On the importance of authorship in pilgrimage landscapes. *Tourism Recreation Research* 27 (2), 27–33.

Altamirano, M.E. (2023) Legitimizing discourses within favela tourism. *Tourism Geographies* 25 (7), 1712–1729.

Barlar, S.H. (2021) Experiences along the Camino de Santiago. In D.H. Olsen and D.J. Timothy (eds) *The Routledge Handbook of Religious and Spiritual Tourism* (pp. 254–265). Routledge.

Beeton, S. (2019) Globalisation, tourism, and pop culture. In D.J. Timothy (ed.) *Handbook of Globalisation and Tourism* (pp. 284–294). Edward Elgar.

Boyd, S.W. (2016) Reflections on slow food: From 'movement' to an emergent research field. In D.J. Timothy (ed.) *Heritage Cuisines: Traditions, Identities and Tourism* (pp. 166–179). Routledge.

Breathnach, T. (2006) Looking for the real me: Locating the self in heritage tourism. *Journal of Heritage Tourism* 1 (2), 100–120.

Cabeça, S.M., Gonçalves, A.R., Marques, J.F. and Tavares, M. (2020) Creative tourism as an inductor of co-creation experiences: The CREATOUR project in the Algarve. In P. Pinto and M. Guerreiro (eds) *Handbook of Research on Resident and Tourist Perspectives on Travel Destinations* (pp. 269–285). IGI Global.

Carvalho, M., Kastenholz, E. and Carneiro, M.J. (2023) Co-creative tourism experiences – a conceptual framework and its application to food and wine tourism. *Tourism Recreation Research* 48 (5), 668–692.

Cetin, G. and Bilgihan, A. (2016) Components of cultural tourists' experiences in destinations. *Current Issues in Tourism* 19 (2), 137–154.

Chambers, E. (2009) From authenticity to significance: Tourism on the frontier of culture and place. *Futures* 41 (6), 353–359.

Chandralal, L., Rindfleish, J. and Valenzuela, F. (2015) An application of travel blog narratives to explore memorable tourism experiences. *Asia Pacific Journal of Tourism Research* 20 (6), 680–693.

Chhabra, D. (2021) Transformative perspectives of tourism: Dialogical perceptiveness. *Journal of Travel and Tourism Marketing* 38 (8), 759–768.

Coghlan, A. and Weiler, B. (2018) Examining transformative processes in volunteer tourism. *Current Issues in Tourism* 21 (5), 567–582.

Collins-Kreiner, N. (2020) A review of research into religion and tourism. *Annals of Tourism Research* 82, Article 102892.

Digance, J. (2006) Religious and secular pilgrimage: Journeys redolent with meaning. In D.J. Timothy and D.H. Olsen (eds) *Tourism, Religion & Spiritual Journeys* (pp. 36–48). Routledge.

Dodds, R. and Butler, R.W. (eds) (2019) *Overtourism: Issues, Realities and Solutions*. De Gruyter.

Dovey, K. and King, R. (2012) Informal urbanism and the taste for slums. *Tourism Geographies* 14 (3), 275–293.

Frisch, T. (2016) Glimpses of another world: The favela as a tourist attraction. In F. Frenzel and K. Koens (eds) *Tourism and Geographies of Inequality: The New Global Slumming Phenomenon* (pp. 126–144). Routledge.

Garau Taberner, J. (2009) Tourist Satisfaction, Dissatisfaction and Place Attachment at Sun and Sand Mass Tourism Destinations. Unpublished doctoral dissertation, University of the Balearic Islands.

Gedecho, E.K. and Nyikana, S. (2023) Religious heritage, tourism, and pilgrimage in Africa. In D.J. Timothy (ed.) *Cultural Heritage and Tourism in Africa* (pp. 85–103). Routledge.

Gheorghilaş, A., Dumbrăveanu, D. and Crăciun, A. (2017) The challenges of the 21st-century museum: Dealing with sophisticated visitors in a sophisticated world. *International Journal of Scientific Management and Tourism* 4, 61–73.

Hall, C.M. and Piggin, R. (2002) Tourism business knowledge of World Heritage Sites: A New Zealand case study. *International Journal of Tourism Research* 4 (5), 401–411.

Harrison, D. and Sharpley, R. (2017) Introduction: Mass tourism in a small world. In D. Harrison and R. Sharpley (eds) *Mass Tourism in a Small World* (pp. 1–14). CABI.

Hernandez-Maskivker, G., Fornells, A., Teixido-Navarro, F. and Pulido, J. (2021) Exploring mass tourism impacts on locals: A comparative analysis between Barcelona and Sevilla. *European Journal of Tourism Research* 29, 2908–2908.

Hopper, P. (2007) *Understanding Cultural Globalization*. Polity.

Iațu, C., Ibănescu, B.C., Stoleriu, O.M. and Munteanu, A. (2018) The WHS designation – A factor of sustainable tourism growth for Romanian rural areas? *Sustainability* 10 (3), Article 626.

Jovicic, D. (2016) Cultural tourism in the context of relations between mass and alternative tourism. *Current Issues in Tourism* 19 (6), 605–612.

Kastenholz, E. and Gronau, W. (2022) Enhancing competences for co-creating appealing and meaningful cultural heritage experiences in tourism. *Journal of Hospitality & Tourism Research* 46 (8), 1519–1544.

Kirillova, K., Lehto, X. and Cai, L. (2017) What triggers transformative tourism experiences? *Tourism Recreation Research* 42 (4), 498–511.

Krings, M. and Simmert, T. (2020) African popular culture enters the global mainstream. *Current History* 119 (817), 182–187.

Lin, Z. and Rasoolimanesh, S.M. (2023) Influencing factors on the intention of sharing heritage tourism experience in social media. *Journal of Hospitality and Tourism Technology* 14 (4), 675–700.

Liutikas, D. (2021) Travel motivations of pilgrims, religious tourists, and spirituality seekers. In D.H. Olsen and D.J. Timothy (eds) *The Routledge handbook of religious and spiritual tourism* (pp. 225–242). Routledge.

Marder, B., Archer-Brown, C., Colliander, J. and Lambert, A. (2019) Vacation posts on Facebook: A model for incidental vicarious travel consumption. *Journal of Travel Research* 58 (6), 1014–1033.

Matteucci, X. (2013) Experiencing flamenco: An examination of a spiritual journey. In S. Filep and P. Pearce (eds) *Tourist Experience and Fulfillment: Insights from Positive Psychology* (pp. 110–126). Routledge.

Matteucci, X. (2022) Existential hapax as tourist embodied transformation. *Tourism Recreation Research* 47 (5–6), 631–635.

Matteucci, X., Koens, K., Calvi, L. and Moretti, S. (2022) Envisioning the futures of cultural tourism. *Futures* 142, Article 103013.

Matteucci, X. and Smith, M.K. (2024) *The Creative Tourist: A Eudaimonic Perspective*. Emerald Publishing.

Mkhize, S.L. and Ivanovic, M. (2020) Building the case for transformative tourism in South Africa. *African Journal of Hospitality, Tourism and Leisure* 9 (4), 717–731.

Olsen, D.H. (2021) Fan pilgrimage, religion, and spirituality. In D.H. Olsen and D.J. Timothy (eds) *The Routledge Handbook of Religious and Spiritual Tourism* (pp. 90–110). Routledge.

Plog, S.C. (1974) Why destination areas rise and fall in popularity. *The Cornell Hotel and Restaurant Administration Quarterly* 14 (4), 55–58.

Podder, A.K., Hakim, S.S. and Bosu, S.P. (2018) Ordinary heritage. *Archnet-IJAR: International Journal of Architectural Research* 12 (2), 334–346.

Poria, Y., Biran, A. and Reichel, A. (2006) Tourist perceptions: Personal vs. non-personal. *Journal of Heritage Tourism* 1 (2), 121–132.

Price, R.H. and Kerr, M.M. (2018) Child's play at war memorials: Insights from a social media debate. *Journal of Heritage Tourism* 13 (2), 167–180.

Pung, J.M. and del Chiappa, G. (2020) An exploratory and qualitative study on the meaning of transformative tourism and its facilitators and inhibitors. *European Journal of Tourism Research* 24, 2404–2404.

Pung, J.M., Gnoth, J. and del Chiappa, G. (2020) Tourist transformation: Towards a conceptual model. *Annals of Tourism Research* 81, Article 102885.

Reiss, S. and Wiltz, J. (2004) Why people watch reality TV. *Media Psychology* 6 (4), 363–378.

Ritzer, G. (2003) The globalization of nothing. *SAIS Review* 23 (2), 189–200.

Ross, D. (2020) Towards meaningful co-creation: A study of creative heritage tourism in Alentejo, Portugal. *Current Issues in Tourism* 23 (22), 2811–2824.

Santos, M.C., Veiga, C. and Águas, P. (2016) Tourism services: Facing the challenge of new tourist profiles. *Worldwide Hospitality and Tourism Themes* 8 (6), 654–669.

Schedneck, B. (2021) Religious and spiritual retreats. In D.H. Olsen and D.J. Timothy (eds) *The Routledge Handbook of Religious and Spiritual Tourism* (pp. 191–203). Routledge.

Seeler, S., Lück, M. and Schänzel, H. (2022) Paradoxes and actualities of off-the-beaten-track tourists. *Journal of Hospitality and Tourism Management* 53, 216–224.

Seeler, S., Schänzel, H. and Lück, M. (2021) Sustainable travel through experienced tourists' desire for eudaemonia and immersion. *Scandinavian Journal of Hospitality and Tourism* 21 (5), 494–513.

Sheldon, P.J. (2020) Designing tourism experiences for inner transformation. *Annals of Tourism Research* 83, Article 102935.

Smith, M.K. (2016) *Issues in Cultural Tourism Studies*. Routledge.

Son, A. and Xu, H. (2013) Religious food as a tourism attraction: The roles of Buddhist temple food in Western tourist experience. *Journal of Heritage Tourism* 8 (2–3), 248–258.

Sugathan, P. and Rakesh Ranjan, K. (2019) Co-creating the tourism experience. *Journal of Business Research* 100, 207–217.

Soulard, J., McGehee, N.G., Stern, M.J. and Lamoureux, K.M. (2021) Transformative tourism: Tourists' drawings, symbols, and narratives of change. *Annals of Tourism Research* 87, Article 103141.

Teoh, M.W., Wang, Y. and Kwek, A. (2021) Conceptualising co-created transformative tourism experiences: A systematic narrative review. *Journal of Hospitality and Tourism Management* 47, 176–189.

Timothy, D.J. (1997) Tourism and the personal heritage experience. *Annals of Tourism Research* 34 (3), 751–754.

Timothy, D.J. (2008) Genealogical mobility: Tourism and the search for a personal past. In D.J. Timothy and J. Kay Guelke (eds) *Geography and Genealogy: Locating Personal Pasts* (pp. 115–135). Ashgate.

Timothy, D.J. (2014a) Contemporary cultural heritage and tourism: Development issues and emerging trends. *Public Archaeology* 13 (3), 30–47.

Timothy, D.J. (2014b) Views of the vernacular: Tourism and heritage of the ordinary. In J. Kaminski, A. Benson and D. Arnold (eds) *Contemporary Issues in Cultural Heritage Tourism* (pp. 32–44). Routledge.

Timothy, D.J. (2018) Making sense of heritage tourism: Research trends in a maturing field of study. *Tourism Management Perspectives* 25, 177–180.

Timothy, D.J. (2019) Globalisation: The shrinking world of tourism. In D.J. Timothy (ed.) *Handbook of Globalisation and Tourism* (pp. 323–332). Edward Elgar.

Timothy, D.J. (2020) Heritage consumption, new tourism and the experience economy. In M. Gravari-Barbas (ed.) *A Research Agenda for Heritage Tourism* (pp. 203–217). Edward Elgar.

Timothy, D.J. (2021a) *Cultural Heritage and Tourism: An Introduction* (2nd edn). Channel View Publications.

Timothy, D.J. (2021b) Heritage and tourism: Alternative perspectives from South Asia. *South Asian Journal of Tourism and Hospitality* 1 (1), 35–57.

Timothy, D.J. and Teye, V.B. (2004) American children of the African diaspora: Journeys to the motherland. In T. Coles and D.J. Timothy (eds) *Tourism, Diasporas and Space* (pp. 111–123). Routledge.

United Nations World Tourism Organization (2018) *Tourism and Culture Synergies*. UNWTO.

UN Tourism (2024) International tourism to reach pre-pandemic levels in 2024. See www.unwto.org/news/international-tourism-to-reach-pre-pandemic-levels-in-2024 (accessed January 2024).
Urry, J. (2016) *What is the Future?* Polity Press.
Vidickienė, D., Vilkė, R. and Gedminaitė-Raudonė, Ž. (2020) Transformative tourism as an innovative tool for rural development. *European Countryside* 12 (3), 277–291.
Wang, N. (1999) Rethinking authenticity in tourism experience. *Annals of Tourism Research* 26 (2), 349–370.
Weiler, B. and Black, R. (2022) The changing face of the tour guide: One-way communicator to choreographer to co-creator of the tourist experience. In G.T. Phi and D. Dredge (eds) *Critical Issues in Tourism Co-Creation* (pp. 91–105). Routledge.
Yamamura, T. and Seaton, P. (eds) (2020) *Contents Tourism and Pop Culture Fandom: Transnational Tourist Experiences.* Channel View Publications.
Zhao, Y. and Agyeiwaah, E. (2023) Understanding tourists' transformative experience: A systematic literature review. *Journal of Hospitality and Tourism Management* 54, 188–199.

8 Museums of the Future: Cultural Tourism Experiences for Wellbeing and Transformation

Marta Šveb Dragija and Daniela Angelina Jelinčić

Introduction

Experiences are no longer enough, visitors now seek to be personally transformed. They want cultural tourism experiences to guide them in their change towards being the best, most authentic, version of themselves (Pine & Gilmore, 2013). The transformative potential of travel has been well documented (e.g. Kirillova et al., 2017; Reisinger, 2013). Travel takes people into unfamiliar situations and places, which allows them to try out new ways of being to satisfy their needs, yielding personal transformation (Kirillova et al., 2017). Ross (2010: 54) suggests that travel 'when approached consciously, can be a widely available, individually tailored, and enjoyable way to gain self-awareness, spiritual experience, and an expansion of consciousness'. UNWTO (2016: 8) equally reports on the transformative role of tourism as having the 'potential to set new paradigms of thinking, to encourage social and cultural changes and to inspire a more sustainable behavior'. Transformation may well apply both at the level of the local community offering tourism services as well as exhibiting paradigm shifts towards more responsible travellers who care about the destination but at the same time expect exceptional and memorable tourism experiences, which may in fact increase their wellbeing.

While transformation is not an easy process as it requires a physiological change in the neural networks, there are indications that museum tourism experiences may become a polygon for experimentation for new models of cultural tourism. Previous studies have already shown that museums are no longer static places, but are assuming new social and political roles (Desvallées & Mairesse, 2010). Yet, the change does not stop here. Bowen and Giannini (2019) stress the dynamism of physical and virtual life, which forces museums to embrace change by maintaining ties with their real and virtual audiences, thus stressing the digital environment as a new, additional space for museum activities. Furthermore, the study by Aizpuru *et al.* (2021) revealed that audiences expect museums to offer activities related to the social welfare of citizens. Therefore, one may assume that future museum activities will increasingly rely on new technologies, stories and experience design in order to facilitate wellbeing and transformation in visitors. These activities have already appeared (mostly stimulated by the COVID-19 crisis); however, they are expected to increase in numbers in the near future, especially due to UNWTO's 2030 Agenda for Sustainable Development and Sustainable Development Goals, which calls for good health and wellbeing to be prioritised in tourism development. Besides visitor engagement, the use of new technologies, and public involvement in content generation (which are current trends in the field of museum experience design; Vermeeren *et al.*, 2018), museums will need to use distinct wellbeing cues to stimulate visitor wellbeing. Emotions will be the key ingredient in experience design because greater emotional engagement is said to lead to memorability, and only memorable events have the potential to transform visitors through promoting their psychological wellbeing (Pine & Gilmore, 2013). Given the dearth of studies in this field, our aim here is to determine the current and future role of museums in fostering tourist transformation and how such transformation can be supported with different internal (e.g. elicitation of emotions and other wellbeing cues shaping experience design) and external museum factors (e.g. outside support for wellbeing initiatives).

Theoretical Background

As one of the most visited cultural tourism attractions (McKercher, 2004), museums have an important role to play in visitor wellbeing, which has so far been demonstrated by several studies (e.g. Chatterjee & Noble, 2016; Jelinčić & Matečić, 2021; Lawler & Tissot, 2021; Šveb Dragija & Jelinčić, 2022; Thomson *et al.*, 2018). Indeed, as noted by Chatterje and Camic (2015), studies have shown that engaging with museums provides

> positive social experiences, leading to reduced social isolation; opportunities for learning and acquiring new skills; calming experience, leading to decreased anxiety; increased positive emotions, such as optimism, hope, and enjoyment; increased self-esteem and sense of identity; increased inspiration and opportunities for meaning-making; a positive distraction from clinical environments, including hospitals and care homes; and increased communication among families, caregivers, and health professional. (2015: 2)

In other words, museums have the potential to impact visitors' longer-term psychological wellbeing, which, according to Ryff (2013), goes beyond momentarily hedonic happiness to prompt personal transformation in terms of increased self-acceptance, personal growth, purpose in life, positive relationships with others, autonomy and environmental mastery. Because one of the necessary psychological mechanisms for psychological wellbeing to occur is reflection (i.e. thinking about oneself, one's actions and world), the International Council of Museums (ICOM) (2022) now includes 'reflection' in its definition of what a museum is by stating that museums 'operate and communicate ethically, professionally and with the participation of communities, offering varied experiences for education, enjoyment, reflection and knowledge sharing'. This recognition of the role of reflection reiterates the likely impact of museums' experiences on visitor wellbeing. Moreover, Helen Chatterjee, a leading researcher in the field of creative health, has confirmed with multiple studies that museums can have a long-term effect on the wellbeing of people suffering from mental and/or physical health problems. For example, object handling sessions in museums which allow visitors to explore objects by touching them are an important tool for increasing the psychological wellbeing of hospital patients (Chatterjee et al., 2009; Thomson et al., 2018). In addition, due to the great benefits gained from museum visitation, Camic and Chatterjee (2013) proposed the formation of strategic partnerships between museums and local healthcare authorities and healthcare funders to offer museums 'on prescription' referral services. A good example is the London Dulwich Picture Gallery, which has one of the most longstanding 'art of prescription' programs for older adults, and which was developed in partnership with local physicians (Harper & Hamblin, 2010).

As a response to this research trend as well as to the stress caused by the COVID-19 pandemic, museums across the world have started to incorporate health and wellbeing activities into their programs, both for marginalised visitors and for the general public. To name just a few, the Museum of Modern Art in New York and Queensland Art Gallery in Australia have established wellbeing programs for people suffering from dementia and their caregivers. The Louvre Museum in Abu Dhabi boasts Arts for Health and Wellbeing program, which includes wellbeing webinar series, kayaking around the museum and yoga under the dome. Also, the Peabody Essex Museum (UK) has a Being Well initiative that focuses on providing yoga in the galleries and online breathing exercises while observing art. By way of further illustration, the National Gallery in Singapore offers a sensory-friendly space called Calm Room, which allows visitors to disengage and rest, while Getty Museum in Los Angeles enjoys a Mindfulness Program that fosters contemplative visitor experience within the museum. These few examples demonstrate that not only museums are increasingly aware of their restorative potential, but also that healing programs are not limited to marginalised groups, as the general public can also benefit from them.

Although museums have only recently started to design experiential spaces and programs for wellbeing, research in this field has a long history

(Šveb Dragija & Jelinčić, 2022). In the early 1990s, researchers were already interested in researching museums as spaces where visitors could rest their minds from the cognitive and emotional overload of everyday life (Kaplan *et al.*, 1993). Early studies found that four museum properties can propel restoration, namely being away from the everyday environment, the extent of time a visitor spends there, a fascination that presupposes inherent interest without mental effort and compatibility of the museum and visitor's needs and purpose. It was also noted that the museum environment should also promote reflection in order to achieve greater wellbeing benefits (Chryslee, 1995; Kaplan *et al.*, 1993). Later, research foci shifted towards participative activities in museums that sought to promote social interaction, such as explorative activities and object-handling sessions (Binne, 2010; Chatterjee *et al.*, 2009; Fenton, 2013; Thomson *et al.*, 2018; Vogelpoel *et al.*, 2013). Currently, there are two complementary research trends in museums and wellbeing. The former combines arts and nature to achieve greater wellbeing benefits for the visitors and the environment (e.g. Thomson *et al.*, 2020), and the latter centres on museum experiences that foster personal growth in visitors (e.g. Šveb Dragija & Jelinčić, 2022; Thomson *et al.*, 2018). To create museum experiences that would contribute to personal growth, it appears essential to first establish a wellbeing culture within the museum organisation. This means that museum leaders need to first hone their growth leadership mindset and leadership skills before they can stimulate personal growth in employees and visitors (Šveb Dragija & Jelinčić, 2023).

Museum experiences that aim at offering meaning and promoting personal growth belong to the largely under-investigated research stream that focuses on eudaimonic wellbeing in tourism. *Eudaimonic tourism* refers to travel experiences that stir tourists' positive emotions but also provide meaning, a sense of achievement and personal growth (Nawijn & Filep, 2016). Eudaimonic tourism experiences are transformational because they go beyond mere positive emotional experiences (such as hedonic experiences) to offer tourists opportunities for self-expressiveness through activities that enable them to find meaning and realise their potential (Šveb Dragija & Franić, 2023). Such meaningful experiences require some effort, experiences of flow and a sense of self-realisation (Matteucci & Filep, 2017; Waterman, 1993). Sometimes, eudaimonic tourism experiences involve negative emotions, which may trigger self-reflection, and which in turn may lead to self-transformation (Šveb Dragija & Franić, 2023). According to Jelinčić (2020):

> while tourism is usually related to the conditionally called positive emotions, an efficient [...] interpretation may not always convey joy and happiness. Negative emotions (e.g. anger, fear or sadness), if successfully elicited in visitors, prove that storytelling and other experience design techniques have been used with effect. (2020: 222)

A Tripadvisor (2014) report reveals that traveller trends are changing from travellers enjoying food and themselves in tourist destinations, to actively seeking new experiences that promote self-reflection, hence eudaimonia. In the same vein, Jelinčić and Senkić (2019: 50) suggest that 'in the future, (creative) tourism will need to: (1) create attractions which are able to relate to individual personal experiences; (2) use creativity in art therapy tourism programmes; and (3) offer experiences that have the power to transform the visitor, thus leaving memorable traces'. In short, in order to capitalise upon the potential that museums have for promoting visitor wellbeing, it is necessary to understand how transformational and wellbeing effects of museum visitation on tourists can be achieved. To investigate this matter, we conducted one focus group discussion with three different stakeholder groups: museum professionals, tourism agencies and decision-makers. The following three research questions guided the research process:

RQ1. What is the role of museums in promoting wellbeing and what will be their role in the future?
RQ2. Which museum factors, internal and external, are necessary for the successful promotion of tourist transformation and wellbeing?
RQ3. What is the role of emotions in promoting transformation and wellbeing of tourists in museums?

Research Methods

To determine the current and future role of museums in the transformation and wellbeing of tourists and how transformation may be achieved and appraised, we carried out a qualitative study in the Croatian context. We conducted a focus group with 10 participants, which included decision makers from tourism and culture ($N = 3$), museum and interpretation centre professionals ($N = 5$), and travel agency representatives ($N = 2$). Focus group is a qualitative technique, which uses in-depth group interviews with participants that are chosen based on their knowledge on the specific topic that is being studied (Rabiee, 2004).

After thorough desk research, we selected 10 participants who through their work showed prominence and authority in the topic of tourist/visitor wellbeing. The names of the participants are not disclosed; hence, pseudonyms are used to refer to them in the findings section. Participants were approached by telephone, informed about the research study and asked to participate in it for which they expressed interest. We opted for conducting an online focus group via Zoom because participants were in different parts of Croatia. The aim of the study was clearly explained to them and we obtained their consent for recording the conversation. The focus group lasted about two hours. The recording of the focus group was transcribed verbatim in Croatian and then translated into English by a bilingual speaker. Thematic analysis was used to analyse the textual data. As a first step, we read the transcripts multiple times, which enabled us to familiarise ourselves with the data. We then used open coding to record the key observations, which were then grouped into initial codes (Miles & Huberman, 1994). As a subsequent step, we engaged in identifying key themes. Finally, evocative quotations best representing each theme were chosen to illustrate the findings.

Findings

The research findings are presented thematically to answer the three research questions. Firstly, we report on the role of museums and their future in promoting tourist wellbeing and transformation. Subsequently, the internal and external factors for the design of wellbeing and transformative museums experiences are presented. Lastly, we address the role of emotions in fostering wellbeing and transformative museum experiences.

The future of museums

The findings indicate that, according to the Croatian experts, the future of museums will bring three types of transformations: transformation of the museums, transformation of the local communities and transformation of the visitors. In other words, the study participants have emphasised that museums' role in society is becoming increasingly polyvalent. Although work in New Museology has already explained the social and political role of museums (Desvallées & Mairesse, 2010), our results show that the role of museums is extending to new fields of activities (e.g. in health generation). At the outset, museums will need to adapt to visitors' needs and wants and develop new functions by focusing on the contemporary topics that are prone to foster personal growth because personal transformation is becoming the goal of most tourist activities. This view is expressed by Ana in the following quote:

> Throughout history, the role of museums have changed, museums must adapt to the current times. The modern role of museums seems to be that museums must be much more active in the local community and more dynamic. Museums need to maintain some basic functions of preservation, heritage collection, but also develop some new functions. This means that they cannot remain only a conservator of the past, but that they need to participate in social reality. Museums must be co-creators of reality.

In addition, museums will have to offer real-life and virtual experiences that promote transformation and wellbeing in tourists. Museums will have to be the place where authenticity can be experienced, the place where visitors cannot experience it anywhere else in the World. Museums will also need to engage in the creation of a virtual museum space where visitors can interact with the museum before, during and after the visit. This virtual space is likely to stimulate longer-term engagement, which may potentially lead to personal transformation. Due to shifting visitor trends, our research participants anticipate virtual spaces to become essential. Indeed, new generations of visitors, who are well-versed in technologies, are also more likely to seek virtual experiences than older generations. To illustrate this point, Sofia stated:

> So, people come to us once or occasionally, and I think that today technology allows us to create a space where we will attract them before and keep them later, where we will enable them to interact, contribute, participate, belong, and thus raise that experience to a higher level. Where we will be able to prepare and influence them even after they have bought a ticket to our museum. A space that is available 24/7 and where they can contribute. I think that should be the future. We must prepare for new generations, for a completely new type of audience. For an audience that consumes content in a completely different way than what we are used to, and I think we will have to adapt to that. The approach with interactivity is not enough for the period ahead.

By designing museum experiences that are attractive, museums can also transform local communities (and tourist offerings). In other words, when museum experiences are designed to appeal to the new wave of travellers seeking entertaining and enriching experiences, museums can significantly boost tourism in the area, resulting in various benefits, such as increased job opportunities and the production of related products and services. One of the ways to achieve this is through opening craft and eco-museums, which are museums that function as interpretation centres for specific areas of interest (e.g. museum of toys, museum of broken relationships, museum of chocolate). These specialised museums would provide several advantages. First and foremost, they would cater to the diverse interests of modern travellers, appealing to those who are looking for unique and intellectually stimulating experiences during their visit. By offering engaging and immersive exhibits, museums would attract a broader range of tourists. Local businesses, such as restaurants, shops and accommodation providers may experience a boost in revenue, ultimately contributing to the economic wellbeing of the community. Furthermore, these museums could stimulate the development of spin-off products and initiatives. For example, a museum dedicated to toys might inspire the production of unique, locally-made toys. At the same time, museums would contribute to enhancing social cohesion by bringing local community members together, united around the same idea. This view is reflected upon by Lucia:

> Ivana's house of fairy tales is the (craft) museum that makes people come to Ogulin; that museum transformed the whole town because it brought people to an otherwise touristically neglected town. The transformation

that this museum has brought to the local community and to generations of children who grew up with its programs is wonderful. It meant a lot to inhabitants of Ogulin, for their creativity and self-awareness.

Internal and external museum factors for transformation and wellbeing

Our analysis indicates that there are both internal and external factors that contribute to museums' capabilities to produce transformative experiences. Internal factors include museum staff, visitors, museum artifacts and exhibits and sustainability aspects around museum experiences. External factors include professional support, public authority support and the interest of the IT sector and other professions in working with museums.

Internal factors

To increase visitors' wellbeing, museum staff will need to hone their interpretation and communication skills, which will be necessary to attract visitors, long before they visit the museum premises, as Philip put it:

> …everyone is looking for some "wow effect" and what you have that would be interesting for them compared to, for example, Amsterdam or Athens. One must not lose one's identity, but tourists really want it to be either some good storytelling, or a short promo tool that conveys attractiveness and is easy to understand.

Furthermore, museum curators and managers should be educated about the variegated ways how to foster visitor wellbeing and transformation. Interdisciplinarity should be cultivated in the museum. It is no longer enough to have educated art curators who 'grow their little garden, a bonsai of museum collection'. Today, as exemplified by Lukas in the following excerpt, museums need interdisciplinary teams of experts who together are able to produce a meaningful interpretive content:

> What we need is absolute interdisciplinarity, we need psychologists, philosophers, filmmakers, playwrights. Interdisciplinarity that will raise all that we have mentioned from emotions and experiences to the level of that ultimately demanding audience that is spoiled more and more by everything. We cannot achieve the "wow effect" by ourselves, even if we are the biggest genius in the world, we still need a team of 10 people to generate the content.

Additionally, to design such museum experiences, it will also be necessary to have growth oriented and motivated museum staff whose ideas are supported by museum directors. This idea is reflected in the words of Jack: 'So, if you have a young employee who is alive and kicking and starts something interesting and if he/she is lucky enough to have a director who recognises it, then something can happen'. For transformational effects, museum visitors will need to be active participants. Also, museum experiences will need to be personalised, interactive, creative and should engage all the senses, as put forward by Maria:

> Tourists are looking for transformation and well-being, they don't want to be passive observers but active participants, and in the future, we need to include them interactively. So, don't set up a museum fund that they then observe, but involve them interactively through workshops and programs that are out of the box. With someone who values heritage, it is necessary to value it in an interesting way.

Moreover, museums should enable tourists to learn new skills, which could be achieved through various workshops on authentic topics that are specific to the museum or its location. For instance, Susan stated:

> The real demand of today's guest is that the guest wants to be an active observer, wants to master some new skills, and not just a cultural tourist. Today, the cultural tourist is not the same as (s)he was 5 years ago, it is not the same as it was 10 years ago. Standard tours are passé, something out of the box is required, some experience is required. The workshops offered in terms of sculpture, ceramics, painting and wine, clay and wine are incredibly successful. The market itself, the demand for such products shows us the importance and need for more of such products or experiences. It reveals [the need] to create workshops of this type as part of the museum. So, we are moving towards creative tourism.

Museums should also aim to become a more dynamic space, where tourists can experience various cultural domains. On the dynamism of museum spaces, Mark suggests that:

> connecting several activities within the museum contributes to the dynamics. For example, museums can have a library and a concert hall. So, as a tourist, you can come here and see an exhibition, go to a concert, or read a book.

The content or the themes of museum exhibitions, when possible, should be focused on contemporary topics, which would allow visitors to learn about themselves. When this is possible, but also or especially when it is not, interpretation (e.g. interesting presentation and storytelling) may be the key to transformative museum experiences, an argument that is reflected in Jack's words:

> …the focus of the museum should be contemporary topics, so curators should try to connect them in some way with museums and what they have inside, what they can give to their community and tourists. This is what tourists would come for, and after they return to their homes so that

they can feel that they have really been enriched in that museum, that they have learned something new, that in some way, this can make them at least a little bit better people.

When designing museum experiences for wellbeing and transformation, museums should consider crafting experiences that are sustainable in the long-term; therefore, a long-term strategy is necessary. In other words, museums should design exhibits and programs that can be consistently offered for an extended duration, recognising that such experiences may take time to be valued by visitors. Furthermore, these experiences should be deeply rooted in and reflect unique aspects of the local culture and its community. A sense of authenticity and a sense of connection to the local context would promote meaningful and long-lasting impressions among museum visitors. This view is articulated by Ana in the following quote:

> For something to be authentic, the local population should definitely be involved in the sense of understanding how they see themselves, and not projecting something from the outside, that is, how we would like tourists to see them.

External factors

To maximise the transformative potential of museum visits, museums would need outside support from professionals who could help them to recognise and seize opportunities for creating meaningful experiences. For instance, decision makers would need to provide support through policy frameworks; travel agencies would need to recognise museum experiences as valuable and enriching; and other professionals (e.g. from the IT sector) would need to develop some interest in working in or with museums. The need for outside support is clearly articulated by Ana and Philip in the following two excerpts:

> The role of the profession is to encourage, to show the beauties that often people who work with such a heritage or live in those cities have no distance from and are not aware of whether it is worth or not. The role of the profession in such processes is to be a mirror and to say, "that's wonderful" and help in certain managerial and professional tasks. But it would never be so effective if local people didn't carry it.

> You have to understand that when you are a curator in institutions, in a community, you cannot do it alone. Someone has to bless the program. It depends on the finances; you are not the last link. You may have an idea, but you cannot realize it unless someone above you endorses it. Even if someone above you blesses it, the tourist board and the city must also bless it.

The role of emotions in transformation

The findings show that tourists have diverging motivations in terms of experiencing different emotions. While some tourists may be motivated to visit a certain museum to feel solely positive emotions, some may also be motivated to experience negative emotions (e.g. sadness, grief), which often bring true enrichment and meaning to their life (e.g. visiting a memorial museum). Although emotions are deeply personal, museums may design spaces to facilitate specific emotional reactions (e.g. negative content may stimulate negative emotions, funny content may stimulate happy emotions). This variety of emotions sought after by museum visitors is asserted by Lukas:

> What is important is what emotion that museum product wants to stimulate in the visitor, because every visitor goes on a trip with a certain demand for an answer to his own emotions and when he finds an answer to his emotions, in other words, when he recognizes within that promotional, virtual and digital work something that can respond to his inner need, he then engages with a specific cultural tourism product and it then remains in his memory, which produces a transformation. For example, a visit to a battlefield stimulates the emotion of glory, strength, and power. There are different emotions that each user wants to recognize and incite during his/her travel.

In addition, although some negative emotions may not be desirable for promoting a tourist experience, they are important ingredients in the tourist transformational process and in fostering wellbeing. Instead of avoiding negative emotions, museums should find creative ways to interpret and present difficult topics, as elaborated by Mark:

> In terms of war tourism, you have to work on changing the narrative, on different storytelling, telling the story so that people want to see it, and not removing it from the list, so that experience is transformed from a trauma into something good that tourists will want to engage with and learn from.

Moreover, to foster transformative visitor experiences, emotions should be perceived to be authentic, in that these should reflect the local community's realities. This notion is explained by Susan: 'Because everything that is authentic somehow contains emotions, or collective emotions towards something, it is necessary for the museum to have a connection with the local community'. At the same time, some emotions should be collective emotions or universal emotions; emotions that are common to everyone across the World. The significance of conveying universal emotions is suggested by Sofia when she says:

> It is precisely this emotion that can be globalized, that is, put into a global context. For example, the Museum of Broken Relationships is a familiar experience for all of us if we are from Africa or Europe. When you put that emotion in a global context, then you somehow bring it closer to the

visitors. You don't have to put your collection in a global context, but that emotion must be global and something that is close and common to all of us as humans.

Discussion

The research findings have highlighted the polyvalent role of museums in promoting wellbeing and in transforming the very museums, their visitors as well as local communities. For museum transformation to happen, it seems necessary to mobilise compelling topics, which can be experienced both in real life settings and virtually. This idea is in line with the work of Aizpuru *et al.* (2021) who found that audiences expect welfare activities from museums. Virtual experiences are seen as specific spaces where visitors can engage before, during and after the visit, or even independently of the visit, which offer them opportunities for longer-term engagement, possibly leading to self-transformation. This is in line with a number of recent studies (e.g. Jelinčić & Senkić, 2019; Šveb Dragija & Jelinčić, 2022; Thomson *et al.*, 2018), which assert a trend towards achieving personal growth through museum experiences. The argument that museums of the future should also exist in the virtual sphere (e.g. in the Metaverse), leaving more time for engagement and reflection, is particularly relevant as virtual environments may transform not only visitors/tourists but also the very notion of what tourism is. While real life tourism experiences are usually short-lived, a deeper connection with a museum that would allow for transformation would be hardly achievable. Virtual engagement, however, offers time-wise a limitless engagement where a deeper meaning and reflection could lead to personal growth. This is concordant with Bowen and Giannini's claim (2019) that museums must be attuned to modern changes and embrace the digital environment for serious engagement with their audiences. However, in terms of offering virtual services to tourists, this is contradictory to the UNWTO (1995) definition of tourism, which qualifies tourism as travel away from home for more than 24 hours. This definition does not include virtual tourism experiences, which have gained prominence during the COVID-19 pandemic; thus, the reconceptualisation of what tourism is seems overdue. Our study also demonstrates that local communities may undergo a process of positive transformation through levering authenticity. For instance, by opening craft and eco-museums, which would display their own identity grounded stories, local communities would be put centre stage and benefit from a sense of pride in their heritage.

Based on our findings, the factors necessary for the successful promotion of museum-related transformation and tourist wellbeing entail (a) qualified and motivated staff (knowledge on wellbeing and transformation, good interpretation and communication skills) working in interdisciplinary teams (e.g. involving decision makers, travel agencies, IT sector, psychologists, sociologists, etc.), (b) engaging museum activities (contemporary topics; personalised and authentic experiences promoting active participation, sensory and emotional engagement; skill-based activities), and (c) long-term museum strategic plans as to ensure sustainability. Our findings confirm the findings of previous studies by Binnie (2010), Chatterjee *et al.* (2009), Fenton (2013), Thomson *et al.* (2018) and Vogelpoel *et al.* (2013), which indicate that engaging museum activities are prerequisites for fostering visitor wellbeing and transformation. The argument that museums need leadership support for transformational experiences is concordant with Šveb Dragija and Jelinčić's (2023) study, who argue that when leaders support the development of museum employees, they create a culture of wellbeing, which, in turn, is essential for facilitating personal growth in visitors.

Lastly, the findings of this study stress the need for emotional elicitation in promoting transformation and wellbeing in museum visitors. In particular, our research participants have noted the important role of both positive and negative emotions in promoting transformation and wellbeing, which is in line with the findings of previous studies that have underscored the value of negative emotions in museum experience design (Jelinčić, 2020) and in fomenting transformative, eudaimonic tourism experiences (Šveb Dragija & Franić, 2023).

Conclusion

This chapter has explored the future role of museums in promoting wellbeing from the perspectives of multiple stakeholders. It was found that museums, as physical places, will have a lesser role to play in serving temporary visitors (such as tourists today), but will assume a new role by serving local community members through transformational activities. The stakeholders that we have interviewed believe that visitor wellbeing and transformation will more likely to happen in the virtual-reality space (e.g. Metaverse museums), which will allow visitors to engage with exhibitions through extended periods of time. In Metaverse museum exhibitions, visitors will interact with computer-generated environments and with other like-minded users. These interactions may benefit visitors in various ways, such as by facilitating stress relief, self-realisation, self-actualisation and other wellbeing outcomes.

Three main factors are likely to impact the successful promotion of visitor wellbeing and transformation in museums. These are: qualified and motivated staff, engaging museum activities and long-term museum strategic plans. These factors reveal the creative, restorative and transformational turn in cultural heritage tourism. Thus, we envision new types of museums to emerge: besides the already well- established eco-museums, craft museums, creative museums, transformational and wellbeing museums will come to light. As already seen in some isolated cases

(Harper & Hamblin, 2010), and as discussed in the literature (Chatterjee et al., 2009; Jelinčić & Senkić, 2019; Thomson et al., 2018), such museums may offer focused art therapy programs, which will require longer visitor engagement in real or virtual environments. Experiences of wellbeing and transformation will not come about without some effort, dedication and the full engagement of the senses (Matteucci, 2022). Therefore, the importance of emotions (both positive and negative) in museum experiences seems unquestionable; in particular, the role of authentic and universal emotions will be prominent in future museum experience design. In the future, the museums which will focus on promoting wellbeing and transformation may be referred to as *eudaimonic museums*.

While promoting wellbeing and transformation is currently en vogue, a future challenge, however will be to measure visitor wellbeing and transformation in the museum context. The mere application of visitor satisfaction scales, albeit a ubiquitous approach, is not adequate for appraising transformation. Given that new generations and types of visitors will emerge, research that seeks to renew our understanding of the needs of various museum audiences will be necessary. Indeed, as suggested by Matteucci et al. (2022), if the future of cultural tourism will be heterotopian, in that clusters of rebellious consumers will seek ethical, emancipatory experiences, out-of-the-norm museum experiences will need to be sensitively designed. Furthermore, in the future, we can expect that Metaverse museums will measure the number of engaged virtual tourists, the duration and levels of their engagement as well as their returning visits. Such activities will require long-term planning, collaborative work within interdisciplinary teams, reconceptualisation of the tourism definition alongside innovative museum and tourism policy frameworks. In short, museum operations, strategies and governance structures will need to be changed if museums want to stimulate visitor emotions, and serve as spaces for visitor transformation.

The data presented in this chapter stems from a small number of research participants, which is a limitation of the current study. Although the informants are experts in the fields of museum and cultural tourism, their views narrowly reflect the Croatian context. Future studies may cover different geographical areas with more participants and using alternative research designs. Finally, it is worth noting that future empirical studies should examine the view, expressed by some research participants, that virtual experiences would allow longer-term engagement with museums, which in turn would foster wellbeing and possible transformation in visitors.

References

Aizpuru, I.A., Cuenca, M. and Cuenca, J. (2021) The future of museums. an analysis from the visitors' perspective in the Spanish context. *The Journal of Arts Management Law and Society* 51 (1), 1–17.

Binnie, J. (2010) Does viewing art in the museum reduce anxiety and improve wellbeing? *Museums & Social Issues* 5 (2), 191–201.
Bowen, J.P. and Giannini, T. (2019) The digital future for museums. In T. Giannini and J. Bowen (eds) *Museums and Digital Culture* (pp. 551–577). Springer.
Camic, P.M. and Chatterjee, H.J. (2013) Museums and art galleries as partners for public health interventions. *Perspectives in Public Health* 133 (1), 66–71.
Chatterjee, H. and Noble, G. (2016) *Museums, Health and Well-Being*. Routledge.
Chatterjee, H., Vreeland, S. and Noble, G. (2009) Museopathy: Exploring the healing potential of handling museum objects. *Museum and Society* 7 (3), 164–177.
Chryslee, G.J. (1995) Creating museums that change people's lives: Operationalizing the notion of restorative environments. *Journal of Museum Education* 20 (1), 17–23.
Desvallées, A. and Mairesse, F. (eds) (2010) *Key Concepts of Museology*. Armand Colin.
Fenton, H. (2013) Museums, participatory arts activities and wellbeing. *Teach. Lifelong Learn* 5, 5–12.
Harper, S. and Hamblin, K. (2010) *This is Living*. Dulwich Picture Gallery.
ICOM (2022) *Museum Definition*. See: https://icom.museum/en/resources/standards-guidelines/museum-definition/ (accessed April 2023).
Jelinčić, D.A. (2020) When heritage speaks t-emoticons: emotional experience design in heritage tourism. In M. Gravari-Barbas (ed.) *A Research Agenda for Heritage Tourism* (pp. 219–234). Edward Elgar Publishing.
Jelinčić, D.A. and Senkić, M. (2019) The value of experience in culture and tourism: The power of emotions. In N. Duxbury and G. Richards (eds) *A Research Agenda for Creative Tourism* (pp. 41–53). Egward Elgar Publishing.
Jelinčić, D.A. and Matečić, I. (2021) Broken but well: Healing dimensions of cultural tourism experiences. *Sustainability* 13 (2), Article 966.
Kaplan, S., Bardwell, L.V. and Slakter, D.B. (1993) The museum as a restorative environment. *Environment and Behavior* 25 (6), 725–742.
Kirillova, K., Lehto, X. and Cai, L. (2017) Tourism and existential transformation: An empirical investigation. *Journal of Travel Research* 56 (5), 638–650.
Lawler, N. and Tissot, A. (2021) Preserving the intangible and immeasurable: Exploring wellbeing frameworks in the museum context. *Journal of the Institute of Conservation* 44 (3), 248–259.
Matteucci, X. (2022) Existential hapax as tourist embodied transformation. *Tourism Recreation Research* 47 (5–6), 631–635.
Matteucci, X. and Filep, S. (2017) Eudaimonic tourist experiences: The case of flamenco. *Leisure Studies* 36 (1), 39–52.
Matteucci, X., Koens, K., Calvi, L. and Moretti, S. (2022) Envisioning the futures of cultural tourism. *Futures* 142, Article 103013.
Miles, M.B. and Huberman, A.M. (1994) *Qualitative Data Analysis: An Expanded Sourcebook*. Sage Publications.
Nawijn, J. and Filep, S. (2016) Two directions for future tourist well-being research. *Annals of Tourism Research* 61, 221–223.
Park, S. and Ahn, D. (2022) Seeking pleasure or meaning? The different impacts of hedonic and eudaimonic tourism happiness on tourists' life satisfaction. *International Journal of Environmental Research and Public Health* 19 (3), 1–15.
Pine, B.J. and Gilmore, J.H. (2013) The experience economy: Past, present and future. In J. Sundbo and F. Sørensen (eds) *Handbook on the Experience Economy* (pp. 21–44). Edward Elgar.
Rabiee, F. (2004) Focus-group interview and data analysis. *Proceedings of the Nutrition Society* 63 (4), 655–660.
Reisinger, Y. (ed.) (2013) *Transformational Tourism: Tourist Perspectives*. CABI.
Ross, S.L. (2010) Transformative travel: An enjoyable way to foster radical change. *ReVision* 32 (1), 54–61.

Ryff, C.D. (2013) Psychological well-being revisited: Advances in the science and practice of eudaimonia. *Psychotherapy and Psychosomatics* 83 (1), 10–28.

Šveb Dragija, M. and Jelinčić, D.A. (2022) Can museums help visitors thrive? Review of studies on psychological wellbeing in museums. *Behavioral Sciences* 12 (11), Article 458.

Šveb Dragija, M. and Jelinčić, D.A. (2023) The art of wellbeing: Determining the leadership mindset for wellbeing in museums. Paper presented at the INTERCOM conference 'The Future Museum. Framing the Skills and Mindsets of the Visionary Leader', 6–9 May 2023, Doha, Qatar.

Šveb Dragija, M. and Franić, S. (2023) Eudaimonic tourism: Ensuring sustainability of tourism through of meaningful tourism experiences. Paper presented at the ToSEE conference 'Engagement & empowerment: A path toward sustainable tourism', 25–27 May 2023, Opatija, Croatia.

Thomson, L.J., Lockyer, B., Camic, P.M. and Chatterjee, H.J. (2018) Effects of a museum-based social prescription intervention on quantitative measures of psychological wellbeing in older adults. *Perspectives in Public Health* 138, 28–38.

Thomson, L.J., Morse, N., Elsden, E. and Chatterjee, H.J. (2020) Art, nature and mental health: assessing the biopsychosocial effects of a 'creative green prescription' museum programme involving horticulture, artmaking and collections. *Perspectives in Public Health* 140 (5), 277–285.

Tripadvisor (2016) TripBarometer April 2016: Global Edition. See: www.tripadvisor.com/TripAdvisorInsights/tripbarometer (accessed April 2016)

Vermeeren, A.P., Calvi, L., Sabiescu, A., Trocchianesi, R., Stuedahl, D., Giaccardi, E. and Radice, S. (2018) Future museum experience design: Crowds, ecosystems and novel technologies. In A.P. Vermeeren, L. Calvi and A. Sabiescu (eds) *Museum Experience Design* (pp. 1–16). Springer International Publishing.

Vogelpoel, N., Lewis-Holmes, B., Thomson, L. and Chatterjee, H. (2013) Touching heritage: Community health and wellbeing promotion through sustainable and inclusive volunteer programming in the museums sector. *The International Journal of the Inclusive Museum* 6, 109–119.

World Tourism Organization (UNWTO) (1995) *Collection of Tourism Expenditure Statistics*. UNWTO.

World Tourism Organization (UNWTO) (2016) *Affiliate Members Global Reports, Volume fourteen – The Transformative Power of Tourism: a paradigm shift towards a more responsible traveller*. UNWTO.

World Tourism Organization (UNWTO) (2023) Achieving the Sustainable Development Goals through Tourism – Toolkit of Indicators for Projects (TIPs). UNWTO.

Part 3
Technologies

9 Cultural Tourists of the Future: Envisioning How Digital Technologies Can Shape Heritage Experiences in Europe

Emanuele Mele[1]

Introduction

The growing importance of culture in visitor experiences (Richards, 2020) has brought unique opportunities for tourism destinations as well as challenges, especially for what concerns meeting the preferences of visitor segments (e.g. differentiated according to purposefulness and depth of cultural motivation) (Richards, 2018), while protecting local heritage (UNWTO, 2021). It is estimated that cultural tourism accounts for 40% of all European tourism, with Germany, UK, Italy, France, the Netherlands and Spain representing the largest markets in terms of cultural tourism opportunities (CBI, 2021). Digital technologies are already playing a role across all the phases of cultural tourism experiences – that is, before, during and after the visit – and their importance has been even accentuated since the COVID-19 pandemic, with a demand for virtual access to cultural sites and performances reaching unprecedented levels (UNWTO, 2020).

Travellers use social media platforms to engage at different levels on various topics, and their use is so widespread that it has come to represent a new way for tourists to communicate and evaluate their experiences concerning cultural heritage. This spans from intangible to tangible attractions, like heritage lodging establishments. The latter can also play a role in urban regeneration (Lee & Chhabra, 2015) – an activity that falls within the broader concept of regenerative tourism and hospitality (RTH), a type of tourism that brings a positive contribution to land and culture by restoring and renewing (Luong *et al.*, 2023). Given these premises, understanding how digital technologies can enable and shape the future consumption of heritage tourism is arguably a topic of great interest for academics, who have highlighted the importance of discussing cultural tourist typologies' preferences (Konstantakis *et al.*, 2022), and authenticity in digital objects (Shehade & Stylianou-Lambert, 2020). This information can then be used by educators and practitioners to prepare themselves and others for a plausible future of heritage tourism experiences.

Addressing these topics, the present conceptual chapter aims at envisioning how digital technologies can influence cultural and heritage experiences in Europe, in a utopian future, taking a demand perspective on the subject. Imagining the future cultural tourists arguably requires crossing multiple resources and viewpoints. Consequently, this conceptual analysis is informed by performing data triangulation from (1) scientific articles, (2) practitioners' viewpoints as well as (3) examples from real cases within the intersection of tourism, culture and digital technologies. The resulting information is then condensed and integrated with the six tourist typologies elaborated by Fan *et al.* (2019) and the five cultural tourist segments identified by McKercher and du Cros (2003). The latter have provided an unparalleled framework that has transformed the way tourism organisations understand cultural tourists (McKercher, 2020). The study by Fan *et al.* (2019) provides accurate and updated tourist groups differentiated based on (online and face-to-face) social contact and destination immersion. Consequently, the theoretical combination and re-elaboration of these segments is deemed suitable to envision a cultural tourist typology from a utopian future perspective.

The concept of utopia can be broadly described as thinking about how the future 'could and should be' (Levitas, 2010: 1), freed from the difficulties that tackle everyday reality; it is 'the desire for a better life, caused by a feeling of discontentment towards the society one lives in' (Vieira, 2010: 6). While acknowledging the critics toward the concept of utopia (Isaac, 2015), it is argued here that by isolating 'specific features in our empirical present so as to read them as components of a different system' (Jameson, 2010: 42) in a utopian, 'ideally perfect' future, it is possible to delineate role models for present cultural and heritage tourism stakeholders (Matteucci *et al.*, 2022). In other words, characteristics of present times and identified trends can be transposed (and transformed) into an ideally perfect system to identify inspirational examples, while still being grounded in reality and, thus, plausible.

This chapter provides a theoretical contribution to the body of research that focuses on the future of culture, heritage tourism and digital technology (e.g. Richards, 2020; Zheng *et al.*, 2023), by investigating the forces for change (e.g. mega trends) that can contribute to shaping a cultural tourist typology of the future, envisioning their unique needs and preferences. In addition to this theoretical contribution, this chapter aims to provide content for academics, lecturers and practitioners, who are searching for suggestions and studies that follow an out-of-the-box, thought-provoking look at the future of cultural and heritage tourism. The objective here is to signal to European destination stakeholders which elements they should start considering catering for future cultural tourists.

The chapter is structured as follows. First, a thorough literature review on future oriented, academic publications concerning culture, heritage tourism and digital technologies is provided. Second, the main conceptual piece outlines academic articles that are triangulated with practitioners' viewpoints and cases to conceptualise three cultural tourist typologies of a utopian future. Finally, the conclusion provides some final considerations regarding the identified cultural tourist groups, as well as managerial implications for destination actors in Europe. It has been the decision of the author to delimit the scope of the chapter on the European region because of (1) the importance of cultural tourism in the area (CBI, 2021), (2) the existence of activities involving European heritage revival and digitalisation (e.g. Egberts, 2014), as well as (3) information availability/accessibility (to the author) concerning pioneering activities and research in the chosen field. Consequently, this direction inevitably leads to results that are primarily meant for European destination stakeholders – even though this does not exclude their possible application in similar sociocultural contexts.

The Future of Computer-Mediated Cultural Tourism: A Demand Perspective

Computer-mediated cultural tourism describes the use of networked telecommunication systems to host, enable and enhance cultural experiences (Romiszowski & Mason, 2004), while cultural tourism refers to 'a form of tourism in which visitors engage with heritage, local cultural and creative activities and the everyday cultural practices of host communities [...]' (Matteucci & Von Zumbusch, 2020: 19). Following this definition, in this chapter, cultural and heritage tourism are used as synonyms.

The cultural tourism market is not homogenous, and research indicates that it can be subdivided into five distinct segments depending on the experience sought (from shallow to deep) and the importance that cultural tourism has when choosing a certain destination (from low to high). Thus, according to McKercher and du Cros (2003) both the 'Serendipitous Cultural Tourist' and the 'Purposeful Cultural Tourist' seek a deep cultural tourism experience. Yet, the former gives little importance to cultural tourism when picking a destination, while for the latter it is crucial. Then there are the 'Incidental Cultural Tourist', the 'Casual Cultural Tourist' and the 'Sightseeing Cultural Tourist', all seeking a shallow type of cultural experience. For these types of tourists, the importance of cultural attractions as pull factors ranges from low (for the incidental segment) to high (for the sightseeing segment). Further research on the intersection between ICT and cultural tourism indicates that these motivation-based segments can be associated with specific visiting preferences, like the interest in the classical past for the 'Purposeful Cultural Tourist' (Konstantakis *et al.*, 2022).

In terms of travel planning, research indicates that most prospective visitors to heritage attractions rely heavily on user-generated content (UGC) and, therefore, their perception and engagement is influenced by platforms where UGC is hosted, like Tripadvisor. These digital channels have played an important role so far throughout the tourism experience, with studies suggesting the heavy use of social media before and while being at the destination, up to the moment of using it to share opinions and emotions once the experience has concluded (Gursoy *et al.*, 2022). In this regard, social media are expected to keep influencing tourism and hospitality for the next 10 years (2023–2033), with tech-savvy Millennials driving social media usage growth (Koumelis, 2023). In the future, social media can become even more efficient and timelier via a connection with artificial intelligence (AI) chatbot programs, like ChatGPT, which would create a seamless process for obtaining and sharing information (Gursoy *et al.*, 2023). Factors influencing the intention of sharing heritage tourism experiences on social media include the available resources to share content, social expectations and the memorability of the tourism experience. Furthermore, research indicates that the latter is particularly important for younger visitors, like Millennials (born between 1980 and 1995) (Lin & Rasoolimanesh, 2023), who together with Generation Z (born between 1996–2010) represent the travellers of the future; as such it is important to learn about their preferences and behavior concerning digital technologies. In this regard, Monaco (2018) explains that the vast amount of information available on the web has made these younger generations more conscious and independent than the previous ones. They are more interested in personalisation, self-growth, and the use of digital tools to travel virtually to destinations. Indeed, augmented reality (AR) and virtual reality (VR) applications may create a future in which cultural tourists can engage with virtual artifacts, and tourists with special needs can opt for personalised experiences with inclusive access, where education is mixed with entertainment (see Prodinger & Neuhofer, 2023).

An extensive literature review performed by Zheng *et al.* (2023) on the use of digital technology for cultural tourism reveals that this type of tourism has entered the digital-oriented phase – that is, cultural tourism 4.0 – characterised by AR and VR, online platforms and virtual tourists. In this regard, the authors invite future demand-oriented research to delve into cultural tourist preferences for digital technology, mobile-mediated cultural tourism, and digital tools' implications in visitors' creativity development, wellbeing and heritage revival. The latter refers to the idea of bringing heritage to *life* by reaching large audiences, through the involvement of multimedia channels that would set the environment for a co-created experience (Egberts, 2014). Egberts (2014) brings this concept to European heritage, highlighting the importance of exploring consumer behaviour in terms of experiences and thinking of enjoyable practices that support the creation of meaning while developing visitors' self-image. Heritage revival through advanced digital technology can also contribute to resolving the issues caused by the presence of crowds in locations of cultural relevance. A study by Frey and Briviba (2021) envisions one future of cultural tourism where AR and VR applications are employed to re-create copies of popular, overcrowded cultural attractions in appropriate physical locations. Using advanced digital technologies, visitors would be able to have an immersive historical experience, which combines entertainment and education – thus drawing from the growing trend of edutainment in tourism. These copies, called 'Revived Originals', are not meant to reduce demand at cultural sites, instead, they are designed to increase supply, by offering an interactive experience to those who may seek to escape congested places, have little resources to visit multiple locations, and wish to have a livelier presentation of cultural heritage.

Beyond the digital-oriented phase of cultural tourism (4.0), there is the Industry 5.0, which is a future vision that aims to combine human experts' creativity with the machines' efficiency, intelligence and accuracy; a combination that was not present in the previous phase (Maddikunta *et al.*, 2021). This would occur in an environment where humans, robots and AI applications work together to deliver immersive, highly personalised experiences to visitors and better decision-making for organisations (Orea-Giner *et al.*, 2022). Taking a futuristic Industry 5.0 approach, a study by Orea-Giner *et al.* (2022) examines the future use of AI for experience provided by cultural institutions. Content obtained from two roundtable discussions with cultural tourists and users, as well as experts, highlights that, in the future, museums and other cultural centers can use AI, first, to fulfill functional tasks automatically, such as providing options for accessible experiences, up-to-date information and client support. This can be achieved by using website chatbots, robots and beacons, which can interact with tourists directly or through their smartphones at the time of the visit. Second, AI programs can be designed to deliver a more immersive experience through the integration of emotional components, like informing visitors on the types of sensations that other people had by looking at a painting. Finally, AI can be used to provide a highly personalised experience for cultural tourists, who can get recommendations based on their profile and preferences before getting to the cultural site; once on site, they can receive content based on their behaviour and interests. In sum, for the future of cultural tourism, this implies the automation of functional tasks, the enhancement of visitor experiences through immersion and personalisation.

The impact of digital technologies on heritage authenticity and regeneration

The use of virtual objects and digital applications to enhance or copy heritage in the future inevitably brings to light the issue of authenticity (Shehade & Stylianou-Lambert, 2020). In short, authenticity can be described as either object-based or activity-related. Objective authenticity is an intrinsic property of the artifact and its origin, it is knowledge-based, and it can be evaluated employing an absolute standard (Jin *et al.*, 2020). As part of object-related authenticity, 'symbolic' or 'constructive' authenticity refers to the imaginary and expectations that tourists project on attractions or activities, which allow them to decide whether what they are experiencing is in fact *authentic* (Wang, 1999) – that is, in line with their expectations and preconceived ideas (Reisinger & Steiner, 2006).

Activity-related authenticity is embedded in the person's subjective experience, and it refers to the concept of existential or experiential authenticity. The latter is about finding one's true identity by discovering other cultures and heritage (Antón *et al.*, 2019), while being close to their 'original spirit' in an out-of-the-ordinary environment (González, 2008; Wang, 1999). This perception of authenticity challenges the centrality of material qualities and shifts the attention to the performance of remembering individual and family memories through plausible experiences, which help visitors understand and engage with heritage – thus engendering an *authentic* emotional response (Smith, 2006). This means that, from a visitor's perspective, authenticity can be reconceptualised as a person's psychological bond with heritage and its attributes (Reisinger & Steiner, 2006), which explains tourists' positive experiences with assets that have no object-based authenticity (Jin *et al.*, 2020). Therefore, it is possible to argue that the appropriate use of digital applications does not have a negative effect on experiential authenticity of heritage attractions – thus, representing an enormous benefit for the future of cultural tourism and ICT.

Addressing these dynamics, scholars state that digital re-constructions of tangible, intangible or even destroyed heritage through virtual, augmented or mixed reality applications must represent faithfully the originals or follow closely the available data, otherwise perceived authenticity may be affected. Documentation should also be provided to visitors to inform them about possible gaps in the sources used for the digital re-constructions of the past (Rogers *et al.*, 2018). This accuracy and faithfulness in the representation of heritage can be arguably ascribed to the so-called 'interactive authenticity', which refers to visitors' thoughts and feelings when engaging with digital or physical reconstructions of the past, including tour guides' narratives and alternative interpretations. This subtype of authenticity has been identified as part of experiential authenticity (Jin *et al.*, 2020), and it arguably allows tourists to feel an activity as both entertaining and educational (González, 2008), which entails negotiation between the interpretation(s) offered at the location and visitors' judgement.

The use of digital technologies, like AR and VR, is also among the criteria for the regenerative design of tangible and intangible heritage. The latter focuses on the revitalisation and safeguard of local knowledge and history. Regenerative design consists of a series of practices meant to 'create a positive interaction between built, human, and natural systems for promoting restoration, renovation, and revitalization of the built environment' (Lucchi, 2023: 2). Lucchi (2023) argues that, on the one side, digital technologies help provide key performance indicators in each design stage (e.g. the analysis of *engagement* data from AR technologies). On the other side, these tools create new experiences and products, like the digital copies envisioned by Frey and Briviba (2021), that are appealing to younger generations of cultural tourists.

The literature suggests that digital technologies not only make heritage more interactive, but they also leave room for a multiplicity of interpretations and narratives that give an alternative to those traditional top-down approaches adopted by the 'Authorised Heritage Discourse' (AHD), according to which only a few selected experts can claim authority over heritage, and they act as guardians of its object-based authenticity (Smith, 2006). Digital technologies can challenge the AHD in the future, expanding the number of lay people – in search for authentic *experiences* – that can get in contact with (virtual) heritage attractions, giving them choice of interaction and interpretation, as well as allowing them to easily voice online their opinions and recommendations with other peers (Shehade & Stylianou-Lambert, 2020). This argument is also in line with research showing that digital platforms, like blogs, already provide non-experts with the possibility to share alternative stories concerning heritage and the people who live in contact with it (Mele & Egberts, 2023); a similar dynamic is observed in those video games where players get to live a virtual reconstruction of the past and its heritage (e.g. Mochocki, 2021). These alternative stories can also be shared in virtual environments (e.g. via VR), where users can interact with each other while being physically distant (Correia Laureiro *et al.*, 2020).

A Cultural Tourist Typology of a Utopian Future

The review of the scientific literature suggests a future where digital technologies can make heritage more accessible, engaging and personalised, while preserving cultural sites through the creation of digital copies. Scholars also indicate a diversity in the concept of authenticity, from *object* to *experience*, which can accommodate a successful integration of virtual products as well as multiple interpretations (e.g. Jin *et al.*, 2020). Empirical research shows that cultural tourists are a heterogenous market. While differences are expected to remain in the future, it is argued that their characteristics will vary. This section of the book chapter takes a theoretical approach to delineate a cultural tourist typology in a utopian future, by triangulating the scientific articles presented in the previous section with practitioners' viewpoints, as well as with pioneering tourism cases that represent the forefront of the sector.

Utopias represent imagined systems that stand for a better future for society, characterised by abundance of resources, harmony and equality. They can also be described as 'early realities', which highlights the possibility for their future realisation. Research outlines that utopias are generated through utopian thinking, which is a tool to envision what a utopian future should look like. This has two components: criticising the current state of things and imagining alternatives (Gretzel, 2021). In this chapter, the focus is on the second component (the future), as there is already plenty of literature that highlights problems related to heritage tourism marketing and management (e.g. Frey & Briviba, 2021; Richards, 2018; Zheng *et al.*, 2020). This approach is thus in line with other future-oriented research that discusses tourism utopias (e.g. Matteucci *et al.*, 2022; Yeoman & McMahon-Beatte, 2016).

Practitioners' viewpoints and pioneering cases in computer-mediated cultural tourism

Computer-mediated, museum experiences arguably represent the forefront of innovation in cultural tourism. An online article published on the art platform Frieze (Thorne, 2015) reveals the viewpoints on the future of museums expressed by 10 curators from various countries in the world. These curators foresee that museums in 2040 will become part of people's daily life, co-existing with every transportation system and even available in floating villages. Future technologies will allow audiences to get inspiration from multiple viewpoints, as well as live personalised experiences based on individual preferences in a virtual, 'imaginary museum'. These will even propose surprise factors (elaborated by algorithms) to go beyond visitors' expectations – which is what streaming platforms have been doing for their users with suggested movies and series. One of the curators argues that, by 2040, while artistic *objects* will remain important for museums, the latter will embrace the concept of art as *a means to* get an experience. Additional considerations brought by these practitioners refer to increased inclusivity for marginalised artists and museums as socialised spaces, where empowered visitors can connect with each other through digital technologies.

Pioneering initiatives on the integration of digital technology and cultural experiences include the use of digital platforms to share museum collections and heritage assets, like the platform 'World Heritage Journeys in Europe' (Figure 9.1), which allows people to undertake cultural journeys directly from their computer or smartphone, and the 'Reproduction of Works of Art and Cultural Heritage' initiative, led by the Victoria & Albert

Figure 9.1 World Heritage Journeys of Europe [Homepage online image] – UNESCO World Heritage Centre (2023) https://visitworldheritage.com/en/eu

Museum, in London (UK) (UNESCO, 2023). In this regard, an article published by UNESCO (2023) reports how digital tools can be used to transmit and revitalise intangible cultural heritage, such as media production projects and multimedia formats. AR and VR can enhance the visitor's experience of cultural heritage sites and museums, to even provide personalised, virtual experiences. Furthermore, AI can create cultural content, such as music and artworks, as well as automatic translations for linguistic diversity.

Following the technology megatrend (Buckley *et al.*, 2015), it is foreseeable that in the future there will be an increase in projects aiming to make people explore cultural heritage through AI technology. Examples of pioneering initiatives that go toward this direction include the program 'AI for Cultural Heritage', launched by Microsoft in 2019, which allows people to explore both tangible and intangible heritage, and the 'RePAIR' (Reconstructing the Past: Artificial Intelligence and Robotics meet Cultural Heritage) project from the Italian Institute of Technology (Yan, 2022). The combination of AI and other advanced technologies not only represents great progress for the restoration and digitalisation of heritage (Yan, 2022), but it also sets the basis for future channels through which people can explore and learn about culture in the future.

Three types of cultural tourists

Following the literature outlined above as well as the practitioners' viewpoints and pioneering initiatives combining cultural heritage with digital technology, Figure 9.2 illustrates the three cultural tourist types in a utopian future where information and communication technology (ICT) plays a key role: Cyber Traveller, Enhanced Traveller and Creative Traveller. These cultural tourist types have been elaborated by, first, triangulating information from scientific articles, practitioners' viewpoints and examples of pioneering initiatives in (digital) cultural tourism from

Europe. Afterwards, the resulting information was merged with the tourist groups identified by McKercher and du Cros (2003) and Fan *et al.* (2019). Within the scope of this chapter, it is important to underline that the latter mainly classifies the identified segments according to their social contact (online vs. face-to-face) and level of encapsulation, which refers to the balance between the holiday mode and daily life commitments, with high encapsulation corresponding to a full *immersion* in a liminal realm. These two concepts, together with the two dimensions (culture centrality and experience depth) illustrated in the study by McKercher (2020), are relevant for the understanding of the three future tourist types proposed in this research. These are described as follows.

Cyber traveller

This type of cultural tourist will use digital technology, like VR or virtual tours on the smartphone or laptop, to travel virtually to European destinations and visit cultural attractions. They can immerse themselves in realistic simulations of historical events, cultural festivals, natural wonders and artistic creations. Consequently, ICT acts here as a key enabler, allowing cyber travellers to teleport themselves to the desired location without worrying about travel expenses, time or risks. They will be able to enjoy various cultural experiences at their fingertips. This characteristic taps into the megatrend of *enabling* technologies, outlined in a report published by OECD (2018). Given the virtual nature of their travel experience, their level of encapsulation is relatively low, as at any moment they can turn off their devices and return immediately to their daily routine and duties. On the one hand, this technology affordance implies low(er) levels of cultural involvement, since the authenticity, serendipity, spontaneity and even uncertainties that may occur when meeting other cultures are drastically reduced in favor of full traveller empowerment – who can decide at any time to turn off interactions or impose their pace to the experience. On the other hand, cyber travellers will enjoy a high online presence, with the possibility of interacting, within a safe environment, with a virtually infinite number of other peers who share similar interests. Thus, cyber travellers will be mainly motivated by factors such as curiosity, convenience and safety, which can be satisfied by fully virtual experiences. These characteristics make cyber travellers resemble already existing segments, like sightseeing cultural tourists (McKercher, 2020) as well as daily life controllers or social media addicts (Fan *et al.*, 2019). The latter are tourist segments that are attached to their comfort zones and original social network, leading to a low level of encapsulation and face-to-face contact.

Enhanced traveller

This type of cultural tourist uses digital technology, like AR, smart glasses and mobile applications, to enhance their travel experience at the destination with timely information and content. Using ICT, enhanced travellers can access relevant facts, stories, reviews and recommendations about the places they visit in Europe. They can also see virtual objects, characters or scenes that add value or entertainment to their cultural exploration. This behaviour resembles aspects of 'THE Digital Companion' scenario elaborated by Prodinger and Neuhofer (2023), based on a series of interviews, according to which in the future there will be a tourist segment that uses digital technology, like virtual assistants and AR glasses, to enhance their travel experience by, for example, visualising historical occasions, community gatherings or how buildings would have looked like in the past. Enhanced travellers are thus motivated by learning, discovery and fun; they want to enrich their understanding and appreciation of different cultures and destinations. The physical presence at the destination allows enhanced travellers to reach a medium cultural involvement, which envisages some contact with the host culture and society, although through the mediation of digital technologies. The latter are used for various purposes, including live interpretations of rituals, symbols and translation services, which, in turn, may increase visitors' cultural awareness and interest. This pragmatic use of digital technology to enhance cultural encounters leads to a low-medium interest in *virtual presence*, in favour of learning – while being entertained – about the local culture through the filter of ICT *at the destination*. Like cyber travellers, enhanced travellers also give high importance to cultural attractions as pull factors. Yet, they reach a medium cultural involvement already from the information-search phase (and afterwards at the destination), which places them in-between sightseeing tourists and purposeful cultural tourists. Considering the above-mentioned dynamics, enhanced travellers are placed in between the 'home zone' and the 'away zone', like the diversionary travellers described by Fan *et al.* (2019), resulting in moderate encapsulation. Indeed, while digital technology is used to enhance the experience at the destination, it inevitably brings ties with the original network. Eventually, their interest in ICT as a means to an end – that is, for its practicality and usage – makes enhanced travellers resemble the pragmatic consumers who emerge from past marketing research (e.g. Kımıloğlu *et al.*, 2010).

Creative traveller

This type of cultural tourists is inspired by the 'Purposeful Tourist' and 'Serendipitous Tourist' segments described by McKercher (2020), who opt for deep cultural experiences, even when these are not essential pull factors for the choice of destination (as in the case of serendipitous tourists). However, unlike these segments, creative travellers use various digital tools, including VR and AR devices, to create their own customised travel experiences, which may reflect their personal values and identities. This type also resembles the dual-zone travellers identified by Fan *et al.* (2019), in that they tend to have 'a high social presence in online

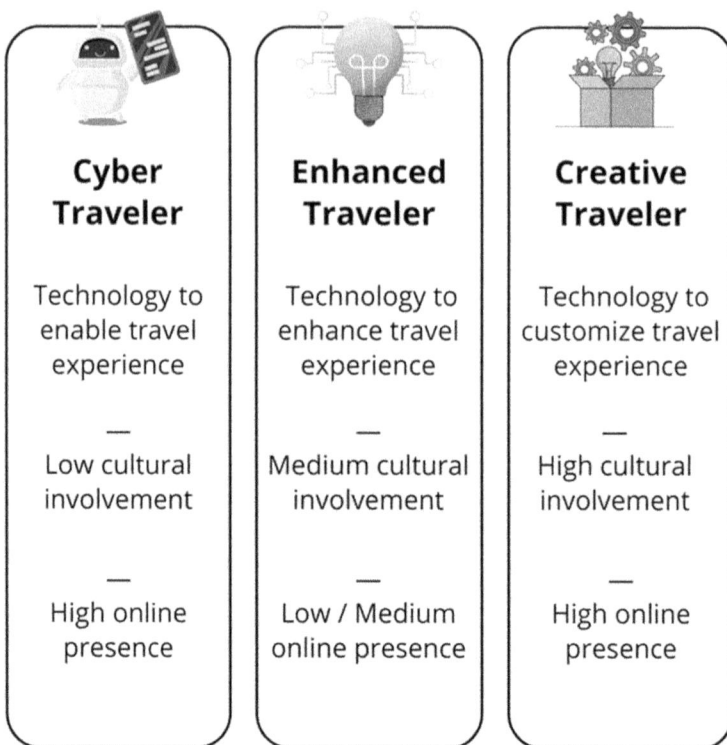

Figure 9.2 Three cultural tourist types of a utopian future

and face-to-face environments' (2019: 10), which means that they try to maximise the connectivity with their original social network (at home) and with other interlocutors at the destination both off- and on-line. Creative travellers can be considered as 'creators of worlds', as they use their deep knowledge and interest in ICT to have the most intense and transformative cultural experience possible. This includes the digital re-creation of popular cultural attractions with the virtual participation of local tour guides. The virtual reconstruction of tangible, intangible and natural heritage is not meant to be exclusively a copy of something that exists or existed in the past. On the contrary, creative travellers can insert their memories and knowledge in what they visualise to display their understanding of European heritage and then discuss it with the local guide. This dynamic represents a (future) extension of the already existing creative tourists (see Matteucci & Smith, 2024) and 'co-created heritage', which has already been documented in the literature (e.g. Ross & Saxena, 2019). Finally, creative travellers allow digital technology to track their intentions and behaviour to then receive personalised suggestions and surprise elements during their experience at the destination. This 'symbiotic' relationship between creative travellers and smart technology makes them the future projection of the 'Pioneer' segment already identified in the literature about technological forecasting and social change (Ramírez-Correa et al., 2020).

Conclusion

This chapter has combined information from scientific articles, practitioners' viewpoints and examples of pioneering tourism initiatives to envision a typology of cultural tourists from a utopian future perspective. As such, these (not-yet-existing) visitor groups present different perceptions (e.g. of *what* is cultural tourism), attitudes and behaviors. It is suggested that multiple futures of cultural tourism through ICT may co-exist, rather than having one that does not leave room for diversity. Indeed, one person's utopia can become another person's dystopia (Claeys, 2013), which is a world where diverging voices are not admissible (Isaac, 2015). Thus, Cyber Travellers perceive cultural tourism as a form of entertainment with or through 'the Other' (e.g. Yan & Santos, 2009); Enhanced Travellers see it as an opportunity for learning and discovery; and Creative Travellers understand it in terms of self-expression and transformation. These three groups reflect to varying degrees three social megatrends that are shaping the future of tourism: the predominance of experiences (even mediated by ICT) over possessions, the quest for transformational travel and the importance of authenticity, which can be a matter of perception. Since these three tourist types are imagined in a utopian future, and whether they use ICT to access experiences from home, receive 'edutainment', or express themselves at the destination, they should be attractive segments of consumers to heritage stakeholders.

Following the characteristics of these three tourist types, in a utopian future, destination actors in Europe should do the following to cater for their needs and preferences. Firstly, virtual experiences and digital copies of heritage attractions should be made available by museums and destination management organisations for people who wish to visit them from home. Suggestions should be integrated to provide content based on self-declared preferences and watch history. Secondly, destination- or attraction-specific apps should be created, and these should allow users to input their thoughts, impressions and artistic re-elaborations in a dedicated section for UGC – which would also create a sense of community. Thirdly, AR and VR applications should be made available for on-site visitors who wish to augment or re-imagine a place of cultural significance, allowing users to select content granularity, perspective (e.g. international vs. local), and participation (e.g. the direct creation of new environments via AI). Finally, digital copies of popular attractions can re-distribute tourism flows and protect destination heritage that would otherwise be damaged by excessive visitor numbers. What is common to all these suggestions is the importance of storytelling for user engagement and memorability, as well as the inclusion of multiple *voices* to reflect the diversity that is at the core of the European identity.

Future research should, first, validate this typology of cultural tourists by using Computer Aided Qualitative Data Analysis on textual content published in cultural tourism-related articles and organisation reports. Second, given its limited scope, this chapter has not addressed the implications of ICT (including AI) from a dystopian future perspective, and the use of decentralised technologies, like blockchains, whose analysis is relevant. Finally, it would be interesting to investigate the type of content that would be more suitable for each type of cultural tourists.

Note

(1) Emanuele Mele is also External Research Associate at USI – Università della Svizzera italiana, UNESCO Chair in ICT to develop and promote sustainable tourism in World Heritage Sites, Lugano, Switzerland.

References

Antón, C., Camarero, C., Laguna, M. and Buhalis, D. (2019) Impacts of authenticity, degree of adaptation and cultural contrast on travellers' memorable gastronomy experiences. *Journal of Hospitality Marketing & Management* 28 (7), 743–764

Buckley, R., Gretzel, U., Scott, D., Weaver, D. and Becken, S. (2015) Tourism megatrends. *Tourism Recreation Research* 40 (1), 59–70.

CBI (2021, May 19) *The European Market Potential for Cultural Tourism*. See www.cbi.eu/market-information/tourism/cultural-tourism/market-potential#which-european-countries-offer-most-opportunities-for-cultural-tourism (accessed April 2023).

Claeys, G. (2013) News from somewhere: Enhanced sociability and the composite definition of utopia and dystopia. *History* 98 (330), 145–173.

Correia Laureiro, S.M., Guerreiro, J. and Ali, F. (2020) 20 years of research on virtual reality and augmented reality in tourism context: A text-mining approach. *Tourism Management* 77, 1–21.

Egberts, L. (2014) Introduction to experience, strategies, authenticity and branding. In L. Egberts and K. Bosma (eds) *Companion to European Heritage Revivals* (pp. 11–30). Springer.

Fan, D.X., Buhalis, D. and Lin, B. (2019) A tourist typology of online and face-to-face social contact: Destination immersion and tourism encapsulation/decapsulation. *Annals of Tourism Research* 78, 1–16.

Frey, B.S. and Briviba, A. (2021) A policy proposal to deal with excessive cultural tourism. *European Planning Studies* 29 (4), 601–618.

González, M.V. (2008) Intangible heritage tourism and identity. *Tourism Management* 29 (4), 807–810.

Gretzel, U. (2021) Conceptualizing the smart tourism mindset: Fostering utopian thinking in smart tourism development. *Journal of Smart Tourism* 1 (1), 3–8.

Gursoy, D., Akova, O. and Atsız, O. (2022) Understanding the heritage experience: A content analysis of online reviews of World Heritage Sites in Instabul. *Journal of Tourism and Cultural Change* 20 (3), 311–334.

Gursoy, D., Li, Y. and Song, H. (2023) ChatGPT and the hospitality and tourism industry: An overview of current trends and future research directions. *Journal of Hospitality Marketing & Management* 32 (5), 579–592.

Isaac, R.K. (2015) Every utopia turns into dystopia. *Tourism Management* 51, 329–330.
Jameson, F. (2010) Utopia as method, or the uses of the future. In M.D. Gordin, H. Tilley and G. Prakash (eds) *Utopia/Dystopia. Conditions of Historical Possibility* (pp. 21–44). Princeton University Press.
Jin, L., Xiao, H. and Shen, H. (2020) Experiential authenticity in heritage museums. *Journal of Destination Marketing & Management* 18, 1–11.
Kımıloğlu, H., Nasır, V.A. and Nasır, S. (2010) Discovering behavioral segments in the mobile phone market. *Journal of Consumer Marketing* 27 (5), 401–413.
Konstantakis, M., Christodoulou, Y., Alexandridis, G., Teneketzis, A. and Caridakis, G. (2022) ACUX typology: A harmonisation of cultural-visitor typologies for multi-profile classification. *Digital* 2 (3), 365–378.
Koumelis, T. (2023, August 28) Future Market Insights: influential impact of social media has resulted in significant growth of the hospitality and tourism industry. See www.traveldailynews.asia/reports-surveys/future-market-insights-influential-impact-of-social-media-has-resulted-in-significant-growth-of-the-hospitality-and-tourism-industry/ (accessed June 2023).
Lee, W. and Chhabra, D. (2015) Heritage hotels and historic lodging: Perspectives on experiential marketing and sustainable culture. *Journal of Heritage Tourism* 10 (2), 103–110.
Levitas, R. (2010) Introduction. In R. Levitas, R. Baccolini, J. Fischer and T. Moylan (eds) *The Concept of Utopia* (Vol. 3, pp. 1–9). Peter Lang.
Lin, Z. and Rasoolimanesh, S.M. (2023) Influencing factors on the intention of sharing heritage tourism experience in social media. *Journal of Hospitality and Tourism Technology* 14 (4), 675–700. https://doi.org/10.1108/JHTT-05-2021-0157
Lucchi, E. (2023) Regenerative design of archeological sites: A pedagogical approach to boost environmental sustainability and social engagement. *Sustainability* 15, 1–25.
Luong, V.H., Manthiou, A., Kang, J. and Nguyen, C. (2023) The building blocks of regenerative tourism and hospitality: A text-mining approach. *Current Issues in Tourism* 27 (3), 361–380. https://doi.org/10.1080/13683500.2023.2228974
Maddikunta, P.K., Pham, Q.-V., Prabadevi, B., Deepa, N., Dev, K., Gadekallu, T.R. and Ruby, R.L. (2021) Industry 5.0: A survey on enabling technologies and potential applications. *Journal of Industrial Information Integration* 26, 1–19.
Matteucci, X. and Smith, M.K. (2024) *The Creative Tourist: A Eudaimonic Perspective.* Emerald Publishing.
Matteucci, X. and Von Zumbusch, J. (2020) *Theoretical Framework for Cultural Tourism in Urban and Regional Destinations.* Deliverable D2.1 of the Horizon 2020 project SmartCulTour (GA number 870708). https://doi.org/10.5281/zenodo.4785433
Matteucci, X., Koens, K., Calvi, L. and Moretti, S. (2022) Envisioning the futures of cultural tourism. *Futures* 142, Article 103013.
McKercher, B. (2020) Cultural tourism market: A perspective paper. *Tourism Review* 75 (1), 126–129.
McKercher, B. and du Cros, H. (2003) Testing a cultural tourism typology. *International Journal of Tourism Research* 5 (1), 45–58.
Mele, E. and Egberts, L. (2023) Exploring travel blogs on tourism and landscape heritage. *Journal of Heritage Tourism* 18 (6), 785–806.
Mochocki, M. (2021) Heritage sites and video games: Questions of authenticity and immersion. *Games and Culture* 16 (8), 951–977.
Monaco, S. (2018) Tourism and the new generations: Emerging trends and social implications in Italy. *Journal of Tourism Futures* 4 (1), 7–15.
OECD (2018) *Analysing Megatrends to Better Shape the Future of Tourism.* OECD Publishing.
Orea-Giner, A., Muñoz-Mazón, A., Villacé-Molinero, T. and Fuentes-Moraleda, L. (2022) Cultural tourist and user experience with artificial intelligence: A holistic perspective from the Industry 5.0 approach. *Journal of Tourism Futures* (Ahead-of-print). https://doi.org/10.1108/JTF-04-2022-0115

Prodinger, B. and Neuhofer, B. (2023) Never-ending tourism: Tourism experience scenarios for 2030. In B. Ferrer-Rosell, D. Massimo and K. Berezina (eds) *Information and Communication Technologies in Tourism 2023* (pp. 288–299). Springer.

Ramírez-Correa, P., Grandón, E.E. and Rondán-Cataluña, F.J. (2020) Users segmentation based on the technological readiness adoption index in emerging countries: The case of Chile. *Technological Forecasting and Social Change* 155, Article 120035.

Reisinger, Y. and Steiner, C.J. (2006) Reconceptualizing object authenticity. *Annals of Tourism Research* 33 (1), 65–86.

Richards, G. (2018) Cultural tourism: A review of recent research and trends. *Journal of Hospitality and Tourism Management* 36, 12–21.

Richards, G. (2020) Culture and tourism: Natural partners or reluctant bedfellows? A perspective paper. *Tourism Review* 75 (1), 232–234.

Rogers, J., Schnabel, M.A. and Moleta, T.J. (2018) Future virtual heritage – Techniques. *2018 3rd Digital Heritage International Congress (DigitalHERITAGE) held jointly with 2018 24th International Conference on Virtual Systems & Multimedia (VSMM 2018)* (pp. 1–4). IEEE.

Romiszowski, A. and Mason, R. (2004) Computer-mediated communication. In D. Jonassen and M. Driscoll (eds) *Handbook of Research on Educational Communications and Technology* (pp. 397–431). Routledge.

Ross, D. and Saxena, G. (2019) Participative co-creation of archaeological heritage: Case insights on creative tourism in Alentejo, Portugal. *Annals of Tourism Research* 79, 1–14.

Shehade, M. and Stylianou-Lambert, T. (2020) Revisiting authenticity in the age of digital transformation of cultural tourism. In V. Katsoni and T. Spyriadis (eds) *Cultural and Tourism Innovation in the Digital Era* (pp. 3–16). Springer.

Smith, L. (2006) *Uses of Heritage*. Routledge.

Thorne, S. (2015, November 25) *What is the Future of the Museum?* See www.frieze.com/article/what-is-the-future-of-the-museum (accessed May 2023).

UNESCO (2023, April 20) *Cutting Edge | Protecting and Preserving Cultural Diversity in the Digital Era*. See https://www.unesco.org/en/articles/cutting-edge-protecting-and-preserving-cultural-diversity-digital-era (accessed July 2023).

UNWTO (2020) *#ShareCulture today to #TravelTomorrow*. See https://www.unwto.org/cultural-tourism-covid-19 (accessed June 2023).

UNWTO (2021) *Tourism and Culture*. See https://www.unwto.org/tourism-and-culture (accessed July 2023).

Vieira, F. (2010) The concept of utopia. In G. Claeys (ed.) *The Cambridge Companion to Utopian Literature* (pp. 3–27). Cambridge University Press.

Wang, N. (1999) Rethinking authenticity in tourism experience. *Annals of Tourism Research* 2 (1), 349–370.

Yan, G. and Santos, C.A. (2009) 'CHINA, FOREVER': Tourism discourse and self-orientalism. *Annals of Tourism Research* 36 (2), 295–315.

Yan, Y. (2022, May 19) *How Technology Facilitates Culture Heritage Restoration and Preservation*. See https://amt-lab.org/blog/2022/5/how-can-technologies-help-with-culture-heritages-restoration-and-preservation (accessed April 2023).

Yeoman, I. and McMahon-Beatte, U. (2016) The future of food tourism. *Journal of Tourism Futures* 2 (1), 95–98.

Zheng, D., Huang, C. and Oraltay, B. (2023) Digital cultural tourism: Progress and a proposed framework for future research. *Asia Pacific Journal of Tourism Research* 28 (3), 234–253.

Zheng, D., Liang, Z. and Ritchie, B.W. (2020) Residents' social dilemma in sustainable heritage tourism: The role of social emotion, efficacy beliefs and temporal concerns. *Journal of Sustainable Tourism* 28 (11), 1782–1804.

10 Beyond Digital Prisons: Counterculture and the Hippie Trail Resurrected

Daniel W.M. Wright

Introduction

> The internet didn't kill counterculture – you just won't find it on Instagram
> Busta, 2021: n.p.

As our world becomes even more digital, so does our culture, and it is our culture that often defines us. The question is, what culture do we want to lead us into the future? What culture do we want to avoid? This chapter considers a future in which society is slipping into a controlled state, where governments and organisations use technology to watch and limit people's movement. Society has been warned about the potential of digital technologies being used as a form of control. Orwell's 1984 is often a moniker for such visions with his future predictions being popular across various discourses. Power (2016) recognises how the 'Big Brother' term is often used as a metaphorical warning about the consequences of government employing modern technology to maintain power and control over its people. Power (2016) also notes the apprehension that people have with the exploitation of data by governments, especially as our digital information becomes more accessible and our digital world becomes more complex. Technology today, with the advancements in artificial intelligence and predictive analytics is advancing to a point that it can collect the data of millions of people and process it in real-time, share that information and make real-time decisions, decisions that could have positive and negative consequences for individuals and organisations. Orwell's future predictions are of a dystopian nature, a future government surveillance state, where the populace is controlled. This chapter considers if society is sliding towards such a future. Importantly, if our future societies do become digitally controlled environments for the masses, then what will counter this, what will exist on the peripheries? It is here where the chapter will consider the future of counterculture and travel.

Travel is valuable because of the cultural interactions and communities it can create, and the counterculture community historically has offered a unique form of shared connection and resistance to dominant culture. History has showcased the power of counterculture ideologies and travelling communities. Gypsies, backpackers and nomads are examples of individuals and collectives escaping the norm, seeking alternative travel experiences, living on the edge of mainstream societies. Hippies and the counterculture movement pursued a lifestyle outside the conventional. These groups have played a significant role in changing and defining current social conditions, beliefs, norms and values. The counterculture movements of the past played a significant role in challenging and changing mainstream structures and beliefs, not only for marginalised members of society, but wider society generally. This chapter therefore considers the value and potential of the counterculture movement as a means of freeing future societies from their digital prisons. Future youth travel communities seeking escapism from controlled digital societies. To explore this idea, the chapter considers the concept of culture wars and cancel culture, the increased digitalisation and techno-managed cities, and ultimately the power of counterculture and travel to drive future cultural change.

Counterculture and the Hippies

The counterculture movement, whose origins are attributed to the early mid-20th century, refers to subcultures whose norms and values and or behaviour contrast significantly to mainstream society, and often purposely operate and function in opposition to that of mainstream society (Saglam, 2014). Gramsci (1973) argued that counterculture presented an opportunity for the working class to escape the hegemonically capitalistic dominance. The 1960s saw a wave in anti-establishment movements across the United States and the United Kingdom, gradually spreading across the Western world. This time saw a growing number of youths opposing traditional values held by society who began to engage in protests. Technological advancements played a significant role at the time, the emergence of television, cinema and news radio all acted as platforms in spreading the counterculture movements to a wider audience. Media platforms help spread the distribution of knowledge, ideas and news stories, further driving the counterculture movement, all of which eventually led to cultural change (Widewalls, 2016). Mainstream values have often been influenced by the ideologies that were originally on the margins of society. Significantly, the counterculture movement of the 1960s ensured society confronted challenging subjects and matters at the time (Wright, 2020). Counterculture movements that were driven by youths were questioning and challenging the more traditionalist and historic beliefs and views. Rojo and Harrington (2017) suggested, 'Counter-culture is essential to growth of culture, and while it can be shocking, disruptive, even painful at times, the wise know that the marginalized often lead the body politic toward a stronger equilibrium, a more perfect union'. Past counterculture movements played an important part in shaping our modern-day society, leading to a less conservative approach. Subsequently leading to attitudes (beliefs and behaviours) that are more liberal in some societies.

The counterculture generation of the 1960s and mid-70s was a period of change in identity, family unit, sexuality, dress and the arts. During this period, through resistance and even revolt, the youth rejected social norms and demonstrated their condemnation of racial ethnic and political injustices (Bousalis, 2021). The term hippie is often attributed to the counterculture movement and was coined by 1960s mass media who aimed to label youth who believed they were acting hip by rejecting societal norms (MacFarlane, 2015). As the counterculture movement gradually came to an end alongside the end of the Vietnam war in 1975, the media continued to keep the hippie image alive, even though it often portrays shallow depictions (Bousalis, 2021). Concerningly, media coverage frequently fails to include the range of diverse individuals who shaped the counterculture movement (Raskin, 2017). Though some hippies did not participate in unruly conduct, the media tends to portray most hippies as radicals who partook in deviant behaviour. The media often looked to lay blame and responsibility of turbulent times to hippie youths, but sociologists suggest that a turbulent society created the hippies (Mills, 2000). At a time when society saw high unemployment and dependence on public and state assistance increased, governments deemed all hippies a burden on society (Friedenberg, 2017).

The Culture Wars

The world culture can be understood as an 'umbrella term' that encompasses the social behaviour, institutions and norms found in our human societies. It brings together the tangible and nontangible elements of different social groups, including knowledge, values and beliefs, art and music, laws and customs, capabilities and resources and habits and behaviours of the individuals and collectives within communities. Culture originates from a specific region and or geographical location, and different cultures exist. Individuals understand, learn and acquire culture through the learning processes of socialisation (Eagleton, 2016). Today, like the past, 'culture' is under the spotlight, and being challenged. However, unlike the past, information, knowledge, ideas, and people can interact, move and travel quicker, thus, the potential for cultural interaction and exchange is more common. Additionally, culture is arguably more diluted and even standardised, much to the impact of globalisation. Modernity was driving standardisation and while the premodern world saw societies living and relying on the local rather than national and global support, globalisation has delivered us with many challenges when it comes to understanding culture. A potential consequence pointed to the standardisation process, a subsequent weakening of cultural practices, skills, and knowledge, because the nonstandard is squeezed to the margins of social existence, and can even become obsolete (Eriksen, 2014).

The term ethnocentrism is applied in social science and anthropology, and it refers to the process when individuals apply one's 'own culture or ethnicity' as a frame of reference to judge other cultures, practices, behaviours, beliefs and people, rather than considering and using the standards of the culture involved/being examined (Hammond & Axelrod, 2006). In today's world, ethnocentric views not only occur in person-to-person interaction, but they take place over online digital platforms. The rise of digital media has shone an even more powerful light onto culture and societies and has shifted the culture zones of conflict. The concept of 'culture wars' is ever more present, particularly across the digital landscape. Culture wars refer to the cultural conflict between different social groups where there is disagreement and polarisation in societal values. Ultimately, the battle is for dominance of one's beliefs, values and practices. The battle for cultural dominance is an important topic for examination, as contemporary culture wars play a significant role in social phenomenon, with different social groups aiming to hold power and supremacy through their values and ideologies and thus, aim to steer public opinion and policy in their direction (Hartman, 2015). Alongside this culture power play is the concept of 'cancel culture'. Cancel culture can be understood as a contemporary phrase originating around the 2010s and early 2020s giving reference to a culture where those who are believed to have acted or spoken in an unacceptable manner are ostracised, boycotted or shunned by society (Ealasaid, 2013). As explained by Dudenhoefer (2020), cancel culture takes a 'mob mentality' approach as a means of a modern social justice practice. While cancel culture is still debated and remains in limbo, a common theme is that it involves taking a public stance against an individual or institution for actions considered objectionable or offensive. According to the Cambridge Dictionary (2023: n.p.), cancel culture is a 'way of behaving in a society or group, especially on social media, in which it is common to completely reject and stop supporting someone because they have said or done something that offends you'. Likewise, they suggest that the main argument against cancel culture is the lack of ability for those who have apparently wronged society to apologise and or learn from their mistakes. Furthermore, arguing that cancel culture has its place to call out and remove problematic people from mainstream culture. Suggesting that it is a method to exert control over a world that is becoming more dangerous and less tolerant. Cancel culture allows us to appoint ourselves the arbiters of right and wrong, and the judge and jury because social media allows us to dole out punishment (Cambridge Dictionary, 2023). However, believing one is right could border on a self-righteous form of ethnocentrism.

Conversely, is cancel culture an effective and suitable approach to holding people accountable, or as a form of punishment without a chance for redemption? What form of punishment is acceptable and who decides? More importantly, on what cultural values and morals are such actions of cancelling people or institutions being based on, the mob? The cancel culture and mob mentality approach are evidently dangerous, and history has often proven that people can often find safety in numbers, even if it means that you are wrong and exerting harm on others. People often follow the mob (masses) because it often feels ok at the time, because we are social animals following social norms. It often goes against our innate feelings to go against what is deemed socially acceptable. Psychological studies have proven that 75% of people will go along with and participant in evil acts (Ross, 2017). The following is chilling thought… 'To look in the mirror and say, it's far more likely than not that if I was a German in WWII, I would have been complicit and actively participated in the genocide of six million Jews' (Ross, 2017: n.p.). Culture today continues to be contested, and the war for cultural dominance like the past, endures. Interestingly, what culture is likely to dominate in the future? Which culture will encapsulate a wider collection of individuals across the world? To consider this, it is necessary to consider how we will live in the future.

The Next Age: Post-Digital

In its simplest explanation, the concept of digital culture is the relationship between people and our interactions with technology. Digital culture allows us to consider how digital media and technology have and continue to shape the lives of people, be it social interactions, work and or leisure time. Digital media can relate to a range of tech instruments that ignite our human senses and allow us to interact, communicate and share knowledge. However, are we arriving at an intersection in time where digital technology will not be a tool to support us, but will become more valuable than humans, and will it be used to control us? So much so, that our next dominant culture will be centred around the growth, power, and influence of digital technology over humans. Today, we are often told that we live in 'the digital age', also termed 'the information age'. This is a time defined by humanities significant strides in computer development, and the capturing of digital information. The digital age refers to many fragments of social impact and disruption can be considered from many different angles. For example, new technologies, the impact on our social lives, social-digital communication, management and workplace changes, corporate finance and economics, transport and infrastructure, sport and leisure, art and culture and much more. Essentially, the digital era can be seen as a disruption to the social dynamics of our everyday life, and consequently, it can be explored and contemplated from many perspectives.

Another common phrase is the fourth industrial revolution. According to Gov.UK (2019: n.p.),

> The Fourth Industrial Revolution is of a scale, speed and complexity that is unprecedented. It is characterised by a fusion of technologies – such as artificial intelligence, gene editing and advanced robotics – that is blurring the lines between the physical, digital and biological worlds. It will disrupt nearly every industry in every country, creating new opportunities and challenges for people, places and businesses to which we must respond.

As noted by Vorobiova (2022) the way people play and work has seen dramatic changes over the past half-century, and the tech visions that previously would have been considered as science fiction, are now our reality. Vorobiova (2022: n.p.) suggests: 'It is important to realize that the digital age is not just one monolithic thing - rather it is a sequence of progressive steps', and at present, we are in the centre of the transformation, termed as mid-digital. Table 10.1 offers an overview of the three stages of the digital ages.

According to Goodwin (2016), in a post-digital age (where we are heading) there will be a further blurring of the real and virtual world, adding to the shading nature and complexity of time and space itself, or at least to what we are currently used to. The next generation(s) are being born as truly digital natives, as babies will be instinctively merged to

Table 10.1 Digital ages

Pre-digital	Technological devices had just one function. Media channels were all one-way communication, TV, newspaper, radio and magazines. Though this phase was not too long ago, this period of technology is one that many look on with fond nostalgia. During this phase, retail was still the primary means of obtaining goods and services. While products gradually transitioned to be more digital with encyclopaedias moving online and phone books becoming searchable repositories.
Mid-digital	The mid-digital phase is where we are now. Companies have embraced digital more and more in concept, but they've not yet fully grasped how expectations have changed. While TV cord-cutting continues to grow in popularity, a majority of the population still uses traditional cable services. There continues to be a disconnect where some transit agencies have shifted to card payments, while others still require cash. There is no continuity of experience across the board.
Post-digital	In the post-digital age, digital itself will move into the background. Just like electricity and its impact on business and individual life, digital will also become ubiquitous. In this new digital age, the internet will be available everywhere and things like smart cars and smart homes will be the norm. The concept of restrictions based on location will be a thing of the past. There will be both new freedoms and new challenges to explore in this age, with a population born where digital is just a fact of life.

Source: Vorobiova (2022)

technology, as new digital environments will interact with the real world, with large corporations and governments accessing data on each individual and, not just tracking their digital and real-world movements, but even watching it and manipulating it. However, more concerningly, is what role and place will humans have in the future, one increasingly dominated by AI? According to Harari (2016), smart artificial intelligence is one of the 21st century's most severe threats to humans. Harari predicts that AI could lead to a rise in a useless class of humans due to the increase in jobless and aimless people. Harari suggests that drugs, virtual reality headsets and gaming could be the solution to keeping such useless individuals entertained. Thus, could this be the dawn of the next counterculture revolution?

Surveillance Cities

To understand future counterculture, it is necessary to understand what society will be like in the future. Counterculture, if present in the future, will likely consist of individuals who live on the edge of society, going against mainstream culture. In his well-established book, *Nineteen Eighty-Four*, Orwell used the following statement throughout, Big Brother is watching you (Orwell, 1949). Its purpose, to remind people that they are under constant surveillance by the authorities. A post-digital era is becoming increasingly real, a world that is like the dystopia visions provided by Orwell. Such a position is raised by Bentham (2019), who notes that Britain risks sliding into a surveillance society, worse than the one predicted in George Orwell's 1984. Tony Porter, (the government's surveillance camera commissioner), notes, people could be left 'cupping hands over their mouths', as cameras capable of identifying people by their walk could soon be operating, all of which would erode personal privacy (Bentham, 2019). With the increasing growth of digital connectivity comes the potential for people to become disconnected (cancelled) and even punished for their behaviour and thoughts. It is important to note that it is not the technology's fault (at least not yet as it is not a sentient being), but the humans implementing and controlling the technology. It is those in power who decipher what is good human behaviour within a particular culture. The Chinese social credit system was initially announced in 2014, a process by which The Chinese Communist Party have been constructing a moral ranking system to monitor the behaviour of its people, ranking individuals, companies and organisations based on their 'social credit'. Punishments can be handed out for bad driving, frivolous spending habits, not paying bills on time, too much time on video games, and social media posts. Punishments could include, but not limited to, reduction of internet speed, access to schools for children and the ability to purchase travel tickets (buses, trains and planes). While the role out is still in its infancy and many limitations are still in place, much to the large population and digital algorithms for monitoring, it still points to a controlled environment managed by state authorities (Canales & Mok, 2022). However, these experiences are not limited to China, as showcased by the following.

Currently, around 56% of the world's population, 4.4 billion inhabitants live in cities (urban environments). By 2050, it is suggested the urban population will more than double its current size with nearly 7 of 10 people living in cities (The World Bank, 2023). Subsequently, what is becoming of our cities and urban environments? The concept of the 'smart city' was initially introduced in 1994 (Dameri & Cocchia, 2013) and saw further appearance from 2010 after projects supported by the EU (Jucevicius *et al.*, 2014). There are some uncertainties around the idea of smart cities, and new terms have also been introduced. A common agreement focuses on the introduction of diverse technologies to help achieve more sustainable environments, integrating digital tech with human systems to attain more efficient energy resources (European Commission, 2023). The idea is a holistic attainment of smart living, where technology, government and society can achieve a smart economy, smart mobility, smart environment, smart people, smart living and smart governance (IEEE, 2023). Since COVID-19 there has been increased attention on the concept of 15-minute cities (Khavarian-Garmsir *et al.*, 2023). According to Moreno (who coined the term in 2016), the concept can be understood as places where 'humans and their well-being are the main purpose of urban organization'. The idea is 'to promote sustainability and health by reducing car dependency and increasing physical activity' (Ables, 2023). After the 2015 Paris Climate Change Conference, the concept became a policy agenda for larger cities with the intention to explore the idea of low-carbon metropolises (Khavarian-Garmsir *et al.*, 2023). Gössling (2020) also suggests that the concept was established as a countermeasure to high vehicle congestion in urban environments. The 15-minute city builds on earlier concepts such as smart cities, all of which focus on the idea that people can access all they need within 15-minutes of walking and/or cycling from their home (Moreno *et al.*, 2021). Moreno *et al.* (2021) argue that the COVID-19 pandemic offered an insight into the difficult socioeconomic impacts on cities and their inhabitants. Cities witnessed record numbers of unemployment and social inequality around the world, and if cities continue to endure lockdowns in the future, then the city concept as we know it today needs to be reconsidered.

Technology is currently being employed across the UK, allowing governments the power to monitor people's movement. For example, traffic cameras and video surveillance are increasingly being introduced across London and other cities (and other cities globally). Sadiq Khan (London Mayor at the time of writing) has introduced the Ultra Low Emission Zone (ULEZ) to help clear up London's air. The ULEZ operates 24 hours a day, 7 days a week, every day of the year, except Christmas Day (25

December). 'The zone currently covers all areas within the North and South Circular Roads. The North Circular (A406) and South Circular (A205) roads are not in the zone. Vehicle that do not meet the ULEZ emission standards need to pay a £12.50 daily charge to drive inside the zone. The ULEZ is to expand across all London boroughs from 29 August 2023' (Transport For London, 2023: n.p.). However, there has been much anger towards such implementation, with revolt and vandalism of the cameras by residents (Manning, 2023; Simpson, 2023), as the consequences of such measures are the restrictions people's movement, with negative social and economic impacts. Similarly, the UK governments 'drive in a clear air zone' is being introduced to cities such as Bath, Birmingham, Bradford, Bristol and Sheffield (Gov.UK, 2023). People wishing to enter these clean air zones are required to pay a daily tariff on an official government website. Outside the cities, the UK is introducing digital motorways, which, according to the National Highways (2023: n.p.), 'Digital Roads will harness data, technology and connectivity to improve the way the Strategic Road Network (SRN) is designed, built, operated and used. This will enable safer journeys, faster delivery and an enhanced customer experience for all'. Thus, access by car to cities is increasingly coming at a price, be it daily tariffs or the ability to own a permitted vehicle.

Another management of movement that has been discussed is the concept of personal carbon allowances (PCA), a tool for achieving climate mitigation targets (Nerini *et al.*, 2021). In 2006, David Miliband (the former Labour environment secretary) suggested a radical scheme for tradable personal carbon allowances (PCAs). Everyone would be given a

> carbon allowance to spend on heating, petrol or flying, issued by a central carbon bank. The national allowance would shrink each year to cut personal carbon emissions. The well-off use the most carbon – they have big houses, drive SUVs and fly frequently – while half the population never flies, and many live in flats or small homes. Some 17m households have no car. Carbon credits would be tradable, so those using least could profit by selling some of their allowance, while heavy users would have to buy their spare credits via the carbon bank. (Toynbee, 2021: n.p.)

According to University College London (2021: n.p.):

> In the wake of the global pandemic, researchers argue that changes in behaviour due to the Covid-19 crisis, combined with increased digitalisation and advancements in ICT and artificial intelligence, offer a "perfect storm of opportunity" to finally implement PCA policy. They call for PCAs to be trialled in certain countries that meet criteria for being climate-conscious and technologically advanced.

Again, another example of how individual freedom and movement could be managed and controlled, and come at a cost.

In a post-digital world, technology will be operating in the background. With growing numbers of connected devices, there will be a drive to implement more 5G networks to meet the growing demand for faster and more reliable internet connections. By 2025, 20.1% of global mobile connections should be operating on 5G, and engineers have already initiated experimenting with 6G, which is said to be able to enable a network connection density 10 times greater than 5G (Statista, 2023). Increased data connectivity will have a significant impact on the access and transfer of real time data. The examples provided here are just a few areas highlighting the potential implementation of technologies that could destabilise individual freedom, moving from the individual towards the state (governments).

Future Counterculture Travel: The Second Coming of the Hippie Trail

> In the internet era, true counterculture is difficult to see, and even harder to find - but that doesn't mean it's not there
> Busta, 2021: n.p.

So, what does counterculture look like in the future? Unlike the past, where counterculture individuals could be identified by their clothes, in the future, counterculture will be less interested in being seen. Instead, it will seek 'freedom from the attention economy, the atomization, and the extractive logic of mainstream communication' (Busta, 2021: n.p.). Future counterculture could see individuals seeking to escape mainstream culture. In the ideas presented here, the future of our social structures are not only tech driven but controlled and regulated. Society has become overly reliant on digital technologies, so much so, that AI and robotics have become more important than humans, to the point that humans are no longer sure of their place and purpose in society. Travel and leisure activities are said to be increasingly restricted and individual movement is also more confined and controlled. According to Busta (2021: n.p.) in the future:

> we can imagine collective held physical spaces reclaimed from empty retail or abandoned venues hosting esoteric local scenes, a proliferation of digital gangs in dark forests who hold secrets ideas, and a new desire for scarcity in cultural object – deeper and closer connections made between people even while rejecting the platform's compulsion to like and share.

As Busta suggests, future counterculture ideologies could be based on a desire to detach from the digital world. Future counterculture travellers will consist of individuals and collectives seeking to escape the prison of the digital world. If our digital approach to living starts to limit and control people's ability to travel and entertain themselves in different locations, then the initial waves of rebellion will commence. Such rebellious actions could replicate and mirror those carried out by the hippies of the 60s and 70s. A second coming of youth rebellion in the future towards controlled digital societies.

The 1960s saw a massive rise in tourists heading to European beaches or short holidays. However, a different breed of tourist had alternative plans, traversing Europe, the Middle East and Asia for months on end, yearning the freedom of dead-end jobs. One of the great expressions of the counterculture movement during the mid-1950s to late 1970s, was the hippie trail (Ireland, 2018). A trek that saw tens of thousands of youths endeavouring on a journey of enlightenment, inner peace, a search for meaning, and for some, simply seeking a good time. These youths were known as hippies, the overlanders, young men, and women seeking greater understanding of the world and their place in it (Open Skies, 2023). There was no single specific route to the hippie trail, rather, a journey in the broader sense, focusing on meaning and purpose. The hippie trail became popular as a counterculture movement for the youth and despite its dwindling numbers over times, the trail remains significant, as the attitudes of the youth reflect a rebellious nature towards their society (Pilot Guides, 2023). Like the hippie trail of the 60s, the future could give rise once more to a hippie travel movement. At a time when people are increasingly disillusioned by their place in the world, due to the rise in technology and AI, high levels of unemployment could once more lead to turbulent times. As noted, counterculture has often driven change and reform across mainstream society. For future generations, a hippie trail reborn, a counterculture movement that seeks freedom from the technocratic, AI digital existence could be the start of setting future generations free from what could be viewed as a dystopian Orwellian society. The rebellious act of future youths, travelling outsides the confines of the digital prison, could be the driver that will set the masses free from the confines of a controlled society. 'To be truly countercultural today, in a time of tech hegemony, one has to, above all, betray the platform, which may come in the form of betraying or divesting from your public online self' (Busta, 2021).

Counter-heterotopias could become present, places and spaces for the non-conformists. The heterotopia concept is often considered as sites of resistance (Topinka, 2010; Matteucci *et al.*, 2022). Matteucci *et al.* (2022: 6) offer further insight in their ideas around heterotopia suggesting:

> a heterotopian future would manifest itself as the multiplication of bubbles of ethical consumption and practices within a slowly decaying mainstream neoliberal order. These bubbles of ethical consumption and practices correspond to what we have introduced as pockets of resistance. Pockets of resistance already exist in the form of cooperatives, social enterprises and non-profit businesses.

However, not only pockets of resistance, but spaces of defiance and escapism from the mainstream. More than just locations and places, the future could see large rural spaces outside the digital urban hubs being occupied by counterculture tourist seeking an escape of the controlled cultural world. Rural environments could become the locations for countercultural happenings. Forests, woodlands, coastal areas, farmlands and highlands, forgotten villages and towns. Spaces and places with limited digital comfort which are less frequented by people, could be the birth places of future digital freedom. A digital escapism would arguably involve a lure to the past, free from digital technology, where one is free to wonder, travel and explore without digital voyeuristic payments and monitoring. Much of the challenges for counterculture travel will depend on the level of digital implementation in the future. People could seek places where technology is not present, be it, payments for entry, services for tourists, digital cameras and even drones capturing people's movement. Future travellers could use countermeasures to maintain their privacy, such as the travel methods they use, that is walking, cycling, hitchhiking, taking time to move from one place to another. A detachment from the digital world would see different forms of payment, new forms of cash or the exchange of services and goods between parties. Future counterculture behaviours could be globally present but limited mainly to domestic travel (at least at the start, as international travel could become more difficult if digital means are necessary). None the less, domestic markets for counterculture travel and experiences could allow tourists the spaces for rebellion and escapism. It is here, where activities, desires and pleasures often forbidden, even cancelled are played out, free from digital censoring. Like the past, and today, future counterculture movements will be a marginalised slice of the wider social reality of countries. It will coexist with other social practices, and even those individuals who are often confined to the digital cities, the rule makers, the mass followers of government dictatorship, even they can still be lured into the counterculture world, seeking an escape from mass conformity. The power of mass social change because of countercultural practices in the future will balance on the level of social content in the controlled digital urban environments. If the digital world becomes a place of misery and suffering from humans, where AI and robotics have taken over, then the exposure of counterculture practices to the masses could be the medicine to move to a new era, an age of reckoning. If future humans seek an escape from digital robotic controlled societies, then it could require a human revolution, and the counterculture travellers could be the start of such a movement.

Conclusion

This chapter set out to explore the growing rise of digital cities, and how society is establishing cultural behaviours that could lead to visions like those depicted by George Orwell in his book *Nineteen Eighty-Four*. As societies head towards a post-digital era, technology is increasingly infiltrating our world. Digital technologies from the internet, social platforms, robotics, 6G, artificial intelligence, big data and recognition capabilities are aligning at a time when cultural conflicts also become more complex, elastic and fleeting. Society is being warned about the potential turbulent nature that could come from the impacts of implementing AI technology, and the future value and importance of humans. This chapter has presented evidence to showcase how an Orwellian state could become a reality, with society building what could be seen as digital prisons. Subsequently, how future generations will be controlled and managed so much so that they will have travel restrictions placed upon them, like ideologies and practices being implemented by the Chinese Social Credit System. However, like the past, in the future there will be individuals and collectives that revolt to mass conformity. The counterculture movement of the past was a mixture of different people, from different backgrounds, seeking a more fulfilled and fair society, going against what they deemed were the unjust actions placed upon them by governments and large corporations.

However, if culture in the future is dominated by digital tech, then like the past, counterculture behaviour could be the remedy for social change. Travellers on the fringe of society, coming together, rebelling, seeking an alternative way of life from the dominant culture, penetrating the minds of the youth, could shift the momentum of mass culture, towards a free future for humans. As noted by Wright (2020: 247), 'the power of counterculture lies within its potential to infiltrate the dominant culture in society, and consequently, allowing for culture change...'. If our cities and inhabitants are controlled and restricted in the future then a new wave of counterculture hippies escaping the technocratic realities presented to them could be the medicine needed to drag society to a new cultural reality, one beyond the digital prison.

References

Ables, M. (2023) '15-minute city' planning is on the rise, experts say. Here's what to know. See www.washingtonpost.com/lifestyle/2023/03/03/15-minute-cities-faq/ (accessed June 2023).

Bousalis, R.R. (2021) The Counterculture Generation: Idolization: Idolized, Appropriated, and Misunderstood. *The Councilor: A Journal of the Social Studies* 82 (2), 1–26.

Busta, C. (2021) The internet didn't kill counterculture – you just won't find it on Instagram. See www.documentjournal.com/2021/01/the-internet-didnt-kill-counterculture-you-just-wont-find-it-on-instagram/ (accessed June 2023).

Cambridge Dictionary (2023) *Cancel Culture*. See https://dictionary.cambridge.org/dictionary/english/cancel-culture (accessed June 2023).

Canales, K. and Mok, A. (2022) China's 'social credit' system ranks citizens and punishes them with throttled internet speeds and flight bans if the Communist Party deems them untrustworthy. See www.businessinsider.com/china-social-credit-system-punishments-and-rewards-explained-2018-4?r=US&IR=T (accessed July 2023).

Dameri, R. and Cocchia, A. (2013) Smart city and digital city: Twenty years of terminology evolution (pp. 1–8) X Conference of the Italian Chapter of AIS, ITAIS 2013, Università Commerciale Luigi Bocconi, Milan (Italy).

Dudenhoefer, N. (2020) Is Cancel Culture Effective? See www.ucf.edu/pegasus/is-cancel-culture-effective/ (accessed July 2023).

Eagleton, T. (2016) *Culture*. Yale University Press.

Ealasaid, M. (2013) Feminism: A fourth wave? *Political Insight* 4 (2), 22–25

Eriksen, T.H. (2014) *Globalization* (2nd edn). Bloomsbury.

European Commission (2023) Communication from the commission. Smart cities and communities – European innovation partnership. Brussels. See https://energy.ec.europa.eu/topics/research-and-technology/energy-and-smart-cities_en (accessed June 2023).

Friedenberg, E. (2017) *Anti-American Generation*. Routledge.

Goodwin, T. (2016) The three ages of digital. See https://techcrunch.com/2016/06/23/the-three-ages-of-digital (accessed June 2023).

Gov.UK (2019) Policy paper: Regulation for the Fourth Industrial Revolution. See www.gov.uk/government/publications/regulation-for-the-fourth-industrial-revolution/regulation-for-the-fourth-industrial-revolution (accessed June 2023).

Gov.UK (2023) Drive in a clean air zone. See https://www.gov.uk/clean-air-zones (accessed June 2023).

Gössling, S. (2020) Why cities need to take road space from cars – And how this could be done. *Journal of Urban Design* 25 (4), 443–448.

Gramsci, A. (1973) *Selections from the Prison Notebooks*. Lawrence and Wishart.

Hammond, R.A. and Axelrod, R. (2006) The evolution of ethnocentrism. *Journal of Conflict Resolution* 50 (6), 926–936.

Harari, Y.N. (2016) *Homo Deus: A Brief History of Tomorrow*. Penguin Publishing.

Hartman, A. (2015) *A War for the Soul of America: A History of the Culture Wars*. University of Chicago Press.

IEEE (the Institute of Electrical and Electronics Engineers) (2014s) AI Ethics in Smart Cities: Starting Points for Discussion in Society. See https://smartcities.ieee.org/newsletter/may-2023/ai-ethics-in-smart-cities-starting-points-for-discussion-in-society (accessed June 2023).

Ireland, B. (2018) The hippie trail and the search for enlightenment. See https://blog.oup.com/2018/03/hippie-trail-search-enlightenment/ (accessed July 2023).

Jucevicius, R., Patašienė, I. and Patašius, M. (2014) Digital dimension of smart city: Critical analysis. *Procedia – Social and Behavioral Sciences* 156, 146–150.

Khavarian-Garmsir, A.R., Sharifi, A. and Sadeghi, A. (2023) The 15-minute city: Urban planning and design efforts toward creating sustainable neighborhoods. *Cities* 132, 1–5.

MacFarlane, S. (2015) *The Hippie Narrative: A Literary Perspective on the Counterculture*. McFarland Publishers.

Manning, C. (2023) Revolt against Ulez spreads with more cameras sabotaged ahead of expansion. See https://metro.co.uk/2023/03/09/london-ulez-more-cameras-vandalised-ahead-of-sadiq-khans-expansion-18411752/ (accessed June 2023).

Matteucci, X., Koens, K., Calvi, L. and Moretti, S. (2022) Envisioning the futures of cultural tourism. *Futures* 142, Article 103013.

Mills, C.W. (2000) *The Sociological Imagination* (40th Anniversary edn). Oxford University Press.

Moreno, C., Allam, Z., Chabaud, D., Gall, C. and Pratlong, F. (2021) Introducing the '15-Minute City': Sustainability, resilience and place identity in future post-pandemic cities. *Smart Cities* 4, 93–111.

National Highways (2023) Digital Road. See https://nationalhighways.co.uk/our-work/digital-data-and-technology/digital-roads/#:~:text=Digital%20Roads-,Digital%20Roads,enhanced%20customer%20experience%20for%20all (accessed June 2023).

Nerini, F.F., Fawcett, T., Parag, Y. and Ekins, P. (2021) Personal carbon allowances revisited. *Nature Sustainability* 4, 1025–1031.

Open Skies (2023) The rise and fall of the Hippie Trail. See https://openskiesmagazine.com/the-rise-and-fall-of-the-hippie-trail/ (accessed July 2023).

Pilot Guides (2023) The Hippie Trail: A Counterculture Movement. See www.pilotguides.com/articles/the-hippie-trail/ (accessed July 2023).

Raskin, J. (2017) Beatniks, hippies, yippies, feminists, and the ongoing American counterculture. In S. Belletto (ed.) *The Cambridge Companion to the Beats* (pp. 36–50). Cambridge University Press.

Rojo, J. and Harrington, S. (2017) When Counter-Culture Becomes Culture: Wastedland 2 And Andrew H. Shirley. See www.huffingtonpost.com/-jaime-rojo-steven-harrington/when-counterculture-becom_b_12646436.html. (accessed July 2023).

Ross, P. (2017) Think You Wouldn't Do Evil? Think Again. See https://observer.com/2017/06/mob-mentality-digital-age-twitter/ (accessed July 2023).

Saglam, B.G. (2014) Rocking London: Youth culture as commodity in the Buddha of Suburbia. *The Journal of Popular Culture* 47 (3), 554–570.

Simpson, J. (2023) New Ulez cameras vandalised amid backlash against zone expansion. See www.telegraph.co.uk/news/2023/03/08/new-ulez-cameras-vandalised-amid-backlash-against-zone-expansion/ (accessed June 2023).

Statista (2023) Forecast number of mobile devices worldwide from 2020 to 2025 (in billions). See www.statista.com/statistics/245501/multiple-mobile-device-ownership-worldwide/#:~:text=In%202021%2C%20the%20number%20of,devices%20compared%20to%202020%20levels. (accessed June 2023).

Topinka, R.J. (2010) Foucault, Borges, heterotopia: Producing knowledge in other spaces. *Foucault Studies* 9, 54–70.

Transport For London (2023) ULEZ Expansion 2023. See https://tfl.gov.uk/modes/driving/ultra-low-emission-zone/ulez-expansion-2023 (accessed March 2023).

University College London (2021) Personal carbon allowances could assist climate targets. See www.ucl.ac.uk/news/2021/aug/personal-carbon-allowances-could-assist-climate-targets (accessed June 2023).

Vorobiova, A. (2022) The Digital Age: The Era We All Are Living In. See https://dzone.com/articles/the-digital-age-the-era-we-all-are-living-in-and-d (accessed June 2023).

Widewalls (2016) Counterculture in Society and Art. See www.widewalls.ch/magazine/counterculture (accessed June 2023).

World Bank (2023) Urban Development. See www.worldbank.org/en/topic/urbandevelopment/overview#:~:text=Today%2C%20some%2056%25%20of%20the,people%20will%20live%20in%20cities (accessed June 2023).

Wright, D. (2020) Counterculture and the future of music festivals and events. In V.V. Cuffy, F. Bakas and W.J.L. Coetzee (eds) *Events Tourism Critical Insights and Contemporary Perspectives* (pp. 232–250). Routledge.

11 Cultural Tourism in the Metaverse

Ulrike Gretzel and Eva Sánchez-Amboage

Introduction

When tourists engage in cultural activities or immerse themselves in cultural experiences at a destination, they do so within specific spatio-temporal, socioeconomic and techno-cultural contexts. This implies that technology, cultural institutions and cultural tourism are intricately linked (Carayannis *et al.*, 2018); whether tourists use digital devices during the experience, the cultural offering is produced by or with technology, or expectations and interactions are mediated by technology. Consequently, changes in the techno-cultural fabric profoundly impact cultural tourism, and cultural tourism often serves as a platform for experimentation with new technologies. From photography to the World Wide Web to the participatory culture and heightened mobilities brought about by social media and mobile devices, culture and, thus, cultural tourism co-evolve with technological development.

Recent government and industry mobilisation towards smart tourism (Gretzel *et al.*, 2015), that is tourism that emerges from strategic efforts at destinations to use advanced technologies in a way that creates new, sustainable tourism opportunities and enriches tourism experiences, has accelerated technology integration into diverse aspects of cultural tourism. The notion of 'smart cultural tourism' (Jankova *et al.*, 2023; Katsoni *et al.*, 2017) signifies an acceleration of technology implementation in cultural tourism offerings, and an ever-greater reliance on ever more sophisticated technology. It is therefore not surprising that robots (Wong & Wong, 2024), artificial intelligence (Thiel & Bernhardt, 2023) and other emerging technologies are increasingly adopted by cultural institutions like museums and galleries. One of the latest technological developments that are profoundly impacting cultural tourism pertains to technologies aimed at realising new forms of digital, immersive experiences, commonly summarised under the label 'Metaverse' technologies.

The Metaverse is conceptualised as the next iteration of the internet, namely an immersive internet (Power & Teigland, 2013), with widespread implications for computing, devices and user experiences. Widely understood as a set of technologies and standards that together propel a vision of an internet that reaches beyond the digital, the Metaverse promises to enable immersive experiences and interactions that current internet users have been speculating about and dreaming of for a while. Like earlier versions of the internet, it is a phenomenon whose creation involves many varied and often conflicting ideas and interests, a multitude of actors, diverse approaches and a large assembly of technological puzzle pieces that need to come together to support its realisation. However, there is a strong consensus as far as the end goal of enriching and expanding the online user experience is concerned.

Many building blocks of the Metaverse have existed for a while but recent progress made in the gaming space and the acceleration of digitisation along with changing attitudes during the COVID-19 pandemic (Gretzel et al., 2020) have propelled the Metaverse idea forward. Add to that major investments by technology companies and the result is an idea of new digital experiences and novel forms of value co-creation that has captured the minds of businesses, technology developers, governments, researchers, educators and consumers around the world, including in tourism and hospitality (Buhalis et al., 2022). Because of tourism's intricate relationship with technology (Werthner & Klein, 1999), it is no surprise that the Metaverse and its related technologies have been extensively discussed in the tourism literature. Papers on virtual and augmented reality abound (Guttentag, 2022), and more recent papers have tried to foreshadow the many possible implications of the Metaverse (Buhalis et al., 2023). Expectations regarding the potential of the Metaverse for cultural tourism are particularly high. Especially younger tourists are eager to have more immersive and engaging cultural experiences (Buhalis & Karatay, 2022). As these new generational cohorts increasingly enter the tourism market, they will progressively drive demand in this direction. This chapter, thus, explores current and future opportunities and challenges for cultural tourism in the Metaverse, with a specific focus on museums, galleries and art installations as examples of cultural institutions and venues that are already extensively using technology to redefine their value propositions.

Role of Technology in Shaping Cultural Tourism

According to UN Tourism (2018), technology is instrumental for ensuring the long-term sustainability, profitability, and competitiveness of cultural tourism products. Cultural tourism deploys technology across a wide spectrum of aspects, from managing and governing cultural tourism institutions to preserving cultural heritage, marketing cultural tourism offerings and managing visitors (Sigala, 2005). This digital transformation of cultural tourism has reshaped the role of cultural tourism institutions and providers by transforming cultural offerings, attracting new audiences and facilitating inclusiveness (Roque *et al.*, 2024). As cultural tourism is experiencing a shift from objects to experiences (Garau, 2017), technologies are increasingly deployed to add value to cultural tourism's experiential offerings (Ponsignon & Derbaix, 2020). Especially significant is the role of technology in moving cultural tourism experiences from their traditional educational focus to the center of the experience realm, as defined by Pine and Gilmore (1998), in which educational, entertaining, esthetic and escapist experiential dimensions merge. Technology supports the creation of such optimal experiences by fostering absorption, immersion and different levels and forms of participation.

Richards (2018) acknowledges this turn towards immersive and creative visitor experiences but adds a practice perspective to deepen our understanding of cultural tourism that is important for understanding the role of technology beyond the experiential perspective. Specifically, he posits that cultural tourism practice needs to be conceptualised as the interplay of resources (which forms the basis of the practice), meanings (which are assigned to the practice) and competences (which are derived from the practice). Technology affects all three components of the cultural tourism practice by facilitating the creation, accessibility, representation and consumption of resources, by supporting the interpretation of (shared) meanings, and by enabling processes that help with the building, maintaining and expanding of competencies. Figure 11.1 summarises this central role of technology within the practice of cultural tourism.

Lu *et al.* (2023) illustrate the many technologies that have and continue to be researched for their potential impact on cultural tourism experiences and practices, from websites to social media and smart technology applications. The impacts that have been explored range from increased attention/learning, changes in visit behaviours and the triggering of emotions/enjoyment. More transformative outcomes of technology use like stewardship (Kang & Gretzel, 2012) or mindfulness (Stankov *et al.*, 2022), and the more dynamic relationships between tourists, cultural products and destination spaces that result from technology use in cultural tourism (Garau, 2017), however, remain underexplored.

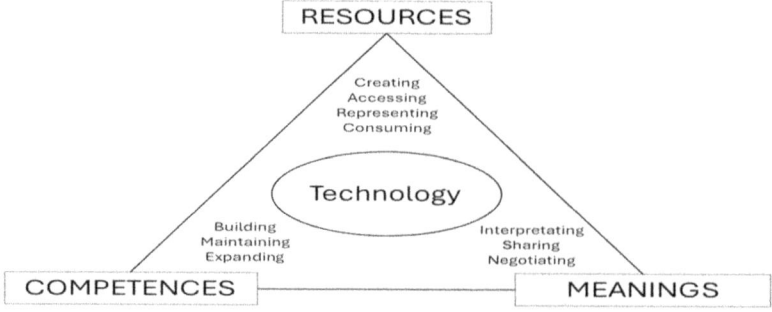

Figure 11.1 The role of technology in shaping cultural tourism practice

In their meta-analysis, Sustacha *et al.* (2023) find that it is the increased informativeness, interactivity, accessibility and personalisation affordances of technologies that affect tourist experiences. While technologies potentially shape experiences across the entire journey of a cultural tourist, affect practices beyond cultural tourism and likely cause spill-over into other life domains given the growing integration of cultural activities into our everyday existence, Hausmann and Schuhbauer (2021) find that technology use differs considerably across visitor segments, which limits the impacts for some cultural tourist segments. In contrast, technological impacts are heightened for those who seek out more participatory offerings. For instance, the creative tourist (Matteucci & Smith, 2024) has been identified as a segment that embraces technologies to create more engaging, embodied and meaningful experiences (Gretzel & Jamal, 2009). Thus, while many cultural tourism institutions are eager to integrate ever more sophisticated technologies into their offerings to remain competitive and relevant, the role of technology in cultural tourism ultimately depends on the cultural tourist's ability and willingness to engage with these technological offerings.

This notion of technological impacts in cultural tourism being driven by both supply and demand is an important one to consider, as the emphasis in academic and industry literature is often on technology adoption and use by the cultural institutions or cultural tourism destinations. For instance, Simone *et al.* (2021) identify four areas of technology use and digital transformation in museums: (1) a back-office use to support cultural heritage preservation mandates; (2) an onsite approach to enhance the experience of museum visitors; (3) an online focus aimed at extending the museum activities and experiences beyond the physical premises; and (4) the 'onlife' transformation of the museum experience through the use of immersive technologies to create accessible, hybrid experiences (Figure 11.2).

However, cultural tourism nowadays happens in smartphone-mediated contexts (Liu *et al.*, 2022) that emerge from consumers travelling with various types of mobile devices that allow them to enrich and expand their immediate physical, social and cultural contexts. Cultural tourism providers can facilitate these technology-mediated experiences or not, but they cannot prevent them. Even if technology use is restricted by a cultural institution, smartphone-enabled tourists will have tapped into official and user-generated digital information before the visit and will enhance their post-experience through available online materials and their own digital storytelling.

While there are many synergies between technology and cultural tourism, there are also areas of potential conflict. Contemporary technoculture, defined as 'the various identities, practices, values, rituals, hierarchies, and other sources and structures of meanings that are influenced, created by, or expressed through technology consumption' (Kozinets, 2019: 621), is fundamentally at odds with the authoritative, conservationist and elitist

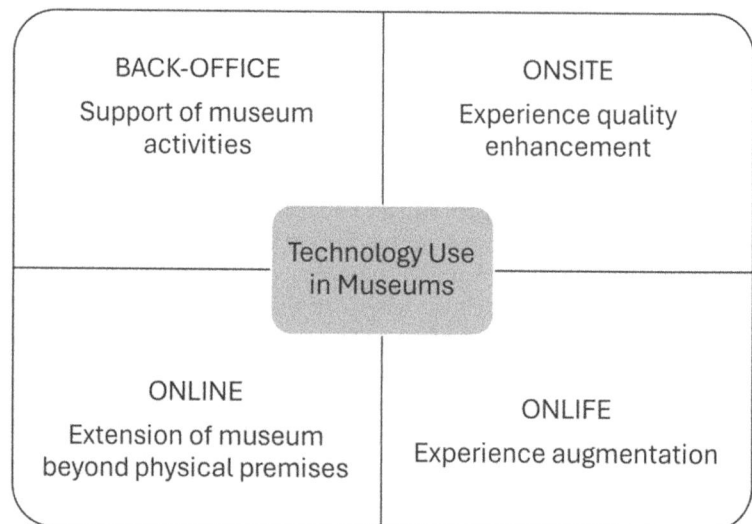

Figure 11.2 Technology use in museums
Source: Adapted from Simone et al., 2021

attitudes that still prevail in many areas of cultural tourism. Contemporary technoculture is inherently participatory (Jenkins, 2006), which resonates with the previous discussion on creative tourists but creates issues for cultural institutions that operate under a different paradigm (Stein, 2012). Roque *et al.* (2024)'s study of Portuguese museums, for example, illustrates that the role of technology in museums often remains limited, and that the notion of a 'participatory museum' continues to be elusive. While the COVID-19 pandemic temporarily woke many cultural institutions from their technology-free sleep, they typically lack technology-focused strategies that would allow them to take advantage of new technological developments (Roque *et al.*, 2024; Simone *et al.*, 2021).

Acknowledging this interplay among technological developments, technology use in everyday life and touristic settings, emerging technoculture and the efforts (or lack thereof) of mainstream players in cultural tourism is important for exploring how the Metaverse and its related technologies and paradigms might impact cultural tourism. To support this exploration, the next section will dive deeper into the concept of the Metaverse and its premises and promises.

Metaverse Building Blocks, Principles and Visions

The Metaverse can be imagined as a digital, immersive and collective space that acts as a sort of layer on top of our lives, or a parallel reality. Mark Zuckerberg famously called it an 'embodied internet you are inside of' (Goldsmith, 2021: n.p.). It does not pertain to a specific technological platform or device, just like the internet does not reside on a specific server and involves many different kinds of websites and applications. Indeed, cross-platforming is a main principle of the Metaverse, which means that it can be envisioned as a vast, interconnected network of 3D-rendered digital environments. According to Techopedia (2024), this interconnectivity is essential to the functioning of the Metaverse, enabling seamless communication, collaboration, and commerce across different virtual environments and platforms. Similarly, Program-Ace.com (2024) describes the Metaverse as 'a comprehensive, interconnected digital universe, amalgamating various forms of reality - virtual, augmented, and physical – into a cohesive, shared space where interactions occur in real time. The persistence of the Metaverse, maintaining its existence irrespective of user presence, mirrors the constancy of the physical universe' (n.p.).

The notion of persistence critically distinguishes the Metaverse from currently available virtual reality and gaming applications. Ball (2022) explains that the Metaverse can be experienced synchronously and persistently by an effectively unlimited number of users with an individual sense of presence, a shared spatial awareness, and with continuity of data, such as identity, history, entitlements, objects, communications and payments. Unlike video games that can be paused and virtual reality applications that start for individual users, the Metaverse does not stop when a user logs off. TechTarget (2024) lists seven technologies as core building blocks for the Metaverse:

(1) artificial intelligence
(2) internet of things
(3) extended reality
(4) brain-computer interfaces
(5) 3D modeling
(6) spatial and edge computing
(7) blockchain.

Artificial intelligence (AI) provides critical support for the computational aspects of the Metaverse as well as its experiential dimensions, especially the creation of content and virtual characters. AI-driven algorithms can enhance user experiences by providing personalised content, natural language processing for communication and realistic behaviours for virtual entities. The Internet of Things will allow for the creation of so-called digital twins (e.g. virtual simulations/representations of physical objects) through networks of sensors. Extended reality (XR) technology encompasses the software, tools and devices that support augmented reality (AR) and virtual reality (VR) experiences. This includes headsets and other wearable devices, such as haptic gloves. Brain-computer interfaces are technologies that translate brain signals into commands that support navigation, avatar-control and communication in virtual environments. They are crucial for the realisation of human digital twins and a human-centric Metaverse (Zhu *et al.*, 2024). So are biometric sensors that capture physiological data from users, such as heart rate, gestures and facial expressions, to personalise experiences and gauge emotional responses within the Metaverse.

3D modelling refers to specialised software needed to create 3D objects and is, thus, a cornerstone technology for the Metaverse. Spatial computing is seen as a critical enabler for the Metaverse as it allows virtual and physical worlds to blend and objects to be manipulated in shared virtual environments. It typically involves a mix of computer vision, sensor fusion and spatial recognition, spatial sound and spatial mapping, often combined with eye-tracking technology that allows for interactions with virtual objects. Edge computing, on the other hand, is a network technology that speeds up processing by moving data processing to the 'edge' of the network, meaning to the device or nearby servers to create efficiency in the network. It is consequently an important element for building the necessary Metaverse computing infrastructure, along with 5G/6G connectivity.

Blockchain technology deserves special attention in the Metaverse context. Blockchain technology is needed in the Metaverse to facilitate security, transparency and decentralisation. Huynh-The *et al.* (2023) see the main blockchain technology applications in the Metaverse as pertaining to the support of data acquisition, data interoperability, data sharing, data storage and data privacy. Many envision the Metaverse to also include a viable economic ecosystem that allows for the creation, ownership and exchange of digital assets. In this context, non-fungible tokens (NFTs) play an important role as they authenticate and serve as a kind of deed or proof of ownership for virtual real estate and virtual artefacts that can be sold or bought in the Metaverse.

Besides the seven technologies listed and described above, interoperability standards and open standards for media are also often mentioned as important for the realisation of the Metaverse. They ensure cross-platform access, asset transfer and identity portability. In essence, they ensure the kind of seamless experience the Metaverse promises. While the development of some of the technologies is moving full steam ahead, it is this lack of interoperability standards and the lack of sufficient computing power that are mostly holding the Metaverse vision back from being fully realised. However, many of the components of the Metaverse are already applied within specific platforms, most prominently within gaming platforms. The question is, therefore not if the Metaverse will happen, but rather when and in what form.

Many aspects of the Metaverse are already being experimented within tourism contexts, and the technological developments connected to the Metaverse will certainly impact tourism in profound ways, just as previous waves of digital innovation have (Werthner, 2022). Buhalis *et al.* (2023) identify the Metaverse as a disruptive technology in tourism because of its ability to create new forms of economic exchanges, facilitate novel types of interactions among players in the tourism system, and its shaping of tourism experiences before, during and after a trip. Koo *et al.* (2023) add that Metaverse tourism will not only afford novel, more immersive forms of touristic experiences, but will also increase the absorptive capacity of tourists for information. Given this promise of enhanced experiences that support highly immersive, spatial, personalised but also shared/social experiences that can involve all four dimensions of the experience realm, possibilities for the applications of the Metaverse or its specific immersive technologies within cultural tourism seem endless.

Cultural Institutions in the Metaverse

Bowen and Giannini (2019) describe contemporary GLAMs (galleries, libraries, archives and museums) as having evolved from highly organised and predictable institutions to 'complex cultural organisms entrenched in the life of diverse communities and connected to global platforms' (2019: 561). This new identity fosters looking for new ways to connect with their stakeholders and to engage a greater array of visitors in new ways. While the COVID-19 pandemic has certainly accelerated their digital transformation, many GLAMs had experimented with immersive technologies quite a while before that (Giannini & Bowen, 2022).

Nevertheless, the adoption of immersive technologies varies significantly across the GLAM landscape. A 2023 survey on technology trends in museums (Johnson, 2023) indicates that the majority of surveyed museums had implemented online exhibitions, projections and visual information displays. However, regarding mixed reality and Metaverse technologies, less than half were planning to deploy smart objects or interactive surface and responsive environments. One third stated that they were planning on introducing augmented and virtual reality, spatial audio and 3D elements. Less than a quarter of the respondents were considering 4D elements, speech interaction, holographic imagery, gesture and motion control or wearable devices. Similarly, Roque *et al.* (2024) found in their study of Portuguese museums that online exhibitions had been implemented by all but one of their cases and two were offering virtual tours, while other immersive technologies were not even on the planning horizon for these museums.

The immersive turn

Giannini and Bowen (2022) identify an additional driver that has pushed GLAMs towards embracing some of the ideas the Metaverse represents, namely the popularity of immersive exhibitions that afford being inside the art. Starting with the birth of so-called 'selfie-museums' in Asia that employed art as props and backgrounds for visitors' selfie-taking quests (Dinhopl & Gretzel, 2016), immersive exhibitions provide technology-backed, multi-sensory, narrative and participatory experiences (Popoli & Derda, 2021). They vary in terms of the setting and technology employed from engaging art displays in public spaces (e.g. airports) and themed entertainment or transformed cultural venues like botanical gardens or castle grounds to travelling museum exhibits like 'Van Gogh: The Immersive Experience'[1], which uses 360-degree displays and VR goggles to immerse visitors in Van Gogh's masterpieces.

Examples of immersive experiences that blend into existing touristic spaces are those created by Moment Factory[2] which include experiences at the Los Angeles International Airport, the Grand Magic Hotel in France, the Sphere in Las Vegas, and the Zoo Sauvage de Saint-Félicien in Canada (among many others). Traditional museums like the Victoria and Albert Museum in London are experimenting with integrating immersive exhibits into their offerings (Giannini & Bowen, 2022). While visitors still move through physical spaces to experience the fascinating art displays that combine visual and sound effects, these immersive exhibitions foreshadow the kinds of interactions with art and cultural heritage that the Metaverse will facilitate. Multi-sensory immersion has significant consequences for the future of GLAM visits as forms of cultural tourism. Using the Mori Digital Museum in Japan[3] as their context, Guo *et al.* (2023) demonstrate that the multi-sensory engagement enabled by such immersive exhibitions has the potential to enhance visitor experiences by positively influencing the emotional state and sense of presence of the museum visitor.

Augmented and virtual reality technologies have been identified as game-changers in the cultural tourism context because of their manifold implications for enhancing visitor experiences, preserving cultural heritage, and extending cultural tourism offerings (Graziano & Privitera, 2020). Lee *et al.* (2020) find that especially the entertainment and esthetic dimensions of a VR experience enhance visitor experiences. The systematic literature review of immersive technologies and tourism-related papers by Pratisto *et al.* (2022) provides many cultural tourism cases, confirming that AR and VR have been, and continue to be, explored extensively in the cultural tourism context, in theory as well as in practice.

An example of AR use in a museum is the extinct animal experience at the National Museum of Natural History in Paris[4] which allows visitors to see animations of extinct animals while wandering through the exhibit. One of the many VR experiences available to visitors is the time-travel experience offered by the Schönbrunn Palace Museum in Vienna[5] which involves visitors sitting in a room with VR goggles. Most museum applications of VR are restricted to wearing a headset; however, more recent advances in VR technology allow for full-body experiences. For example, the Cleveland Museum of Natural History[6] uses VR technology that allows visitors to experience sensations of flight. Thus, while immersive

AR and VR technology applications in cultural institutions currently mostly use visuals and sounds, haptics and proprioception will increasingly be integrated in the future to deliver the kinds of immersive experiences cultural tourists have become accustomed to.

From location-based to remote immersion

Not surprisingly, most GLAMs currently implement immersive technology at their physical premises to enhance the quality of experiences and attract visitors. These integrations of technology into the museum space correspond to Simone *et al.* (2021)'s 'Onsite' category of technology use for experience quality enhancement described in Figure 11.2. Such location-based uses of immersive technologies include holographic models and 3D visualisations, immersive exhibits, AR and VR experiences (Table 11.1).

Holographic displays and 3D visualisations have become common place in many GLAMs. For example, the Daintree Discovery Centre in Queensland, Australia offers a 3D hologram attraction that allows visitors to learn about the culture and country of First Nation people in the area of KuKu Yalanji. There, Juan Walker, a KuKu Yalanji tour guide from Walkabout Cultural Adventures explains the cultural significance of different animals, from the culinary to the mystical, as hologram versions appear before the visitor. Holographic displays have become themselves objects of art that are being collected by the MIT Museum Art Collection[7] and displayed by the Center for the Holographic Arts[8]. The Museum of the Future in Dubai[9] combines holographic displays with an incredible immersive exhibit that takes visitors into the future, where they can experience Dubai in the year 2071. Immersive exhibits have also become a form of art, as exemplified by the 'Infinity Mirror Rooms' exhibit by the immersive art pioneer Yayoi Kusama at The Broad Museum in Los Angeles[10].

Catering to the needs of digital cultural tourists who want personalised, engaging and immersive experiences, AR technology is becoming more common place in GLAMs. For example, the Smithsonian American Art Museum in Washington, D.C. has incorporated AR into its exhibits to enhance the visitor experience. The 'America Now' AR app allows visitors to explore artworks in a new and interactive way. By pointing their smartphones or tablets at specific artworks, visitors can unlock additional

Table 11.1 Typology of immersive technology experiences

Types of Immersive Technology Experiences

Location-based				Remote					
Holographic Model/3D Visualisation	Immersive Exhibit	Augmented Reality	Virtual Reality	Online Exhibit	Virtual Visit	Virtual (360°) tour	Digital Twin	VR Content	Virtual World
Onsite				Online			Onlife		

digital content, such as videos, audio recordings, historical information and interactive elements overlaid on the physical artwork. Location-based VR is also becoming more popular to complement existing displays. An example is the VR experience provided by the museum Banská Štiavnica, a UNESCO World Heritage Site in Slovakia, that allows visitors to journey back in time and experience the medieval town with the help of VR headsets[11]. Zhou *et al.* (2022) reveal that VR and AR have primarily been used in science, art, and history museums to support science and art learning, but both technologies are expected to become more widely adopted by GLAMs.

Itani and Hollebeek (2021) and Sánchez-Amboage *et al.* (2023), among others, explain how the COVID-19 pandemic pushed a great number of GLAMs into explorations of ways to extend their offerings into the digital realm to serve visitors while in-person visits were restricted. Simone *et al.* (2021) refer to these attempts as 'Online' strategies that allow GLAMs to reach and engage visitors beyond their physical locales. These digital innovations have had a lasting impact on cultural tourism because they can now be used in lieu of or in conjunction with physical visits. They include online exhibits, virtual visits and virtual tours (Table 11.1).

Online exhibits are very common today and let digital cultural tourists explore artworks and historic objects from anywhere. An example is the online exhibit of a space suit worn by Helen Sharman featured on the London Science Museum website[12]. Virtual visits became possible courtesy of advances in online conferencing software like Zoom or Teams. Online guided tours or virtual visits are offered by the Thyssen Museum in Madrid for a small fee[13] and can include a live virtual visit with a human guide. In contrast to virtual visits, virtual tours are more interactive. Examples of virtual tours are the Smithonian National Museum of Natural History[14], the virtual gallery visit offered by the Asian Civilisations Museum in Singapore[15], and the various curated tours offered by the Thyssen Museum in Madrid[16]. They let visitors explore the museum offerings at their own pace using technology that has its origins in 360° photography and has been applied in tourism for over two decades (Cho *et al.*, 2002). An interesting extrapolation of this 'online' trend is the Kremer Museum[17], a digital art museum that only exists online.

The GLAM offerings that most closely reflect the ideas of the Metaverse are those that Simone *et al.* (2021) refer to as 'Onlife' experiences, where technology is used to augment experiences beyond traditional GLAM contexts. Table 11.1 shows that these include digital twins, content produced for consumption with personal VR headsets, and content to be experienced in virtual worlds.

Cruz Franco *et al.* (2022) describe applications of digital twin technology for architectural heritage. Also, the National Library of Medicine[18] is creating digital twins of rare books and is using immersive technology to simulate the experience of turning the pages of said books. As far as VR content is concerned, Meta, for example, currently offers a VR museum[19] that can be experienced using its VR headset technology. Examples of 'metaverse-native' museums include the Moonshin Art Museum (Roblox Studio) (Kang *et al.*, 2022) and the Museum: Renaissance (Zepeto World) (Lee *et al.*, 2022). Real-world museums that have created a presence on a virtual world platform are still rare, but they do exist. For instance, the National Museum of Korean Contemporary History is building a virtual museum on Zepeto[20]. Metaverse visitors experience these virtual world exhibits through their avatars and can encounter other visitors in the process.

Related to this engagement with the Metaverse are the forays of GLAMs into the world of NFTs and their support of AI-generated and virtual reality art (Onix, 2024). Jung (2023) points out that the current use of NFTs in the museum sphere is quite limited, and so far, it appears to be adopted only by large and well-known museums. The author highlights two options for museums regarding NFTs: creating NFTs based on museum collections or establishing galleries/museums that sell and showcase their artworks in NFT format. The Museum of Crypto Art[21] is an example of an NFT-focused gallery that exhibits across several virtual worlds as well as in physical locations all over the world. A prominent example of this kind of Metaverse engagement by an established museum is the Museum of Old and New Art (Mona) in Hobart, Australia. Called the 'Monaverse'[22], the museum's dedicated NFT platform features NFT artwork for purchase and exhibits virtual reality art/3D objects created on the platform. Purchased NFT art can then be displayed in NFT galleries on virtual world platforms, suggesting that individual users can become virtual art curators and virtual gallery/museum owners. This clearly reflects the participatory nature of Metaverse-inspired contemporary technoculture.

The Metaverse not only changes how we consume culture, but also profoundly impacts art and culture itself. The growing use of AI for art generation and the use of NFTs to sell and own art has sparked a widespread debate about the role of technology within the artistic community and the meaning of art collections and museums (Hurst *et al.*, 2022). Carayannis *et al.* (2018: 145) stress that as 'art is continuously exploring and experimenting with new forms, this also is (or should be) the case for museums'. In many ways, these debates are linked to broader discussions of the role of museums and cultural institutions in the future (Carayannis *et al.*, 2018) and of the role and meaning of virtual cultural tourism (Guttentag, 2022). These debates will certainly become even more prominent as Metaverse developments progress.

Conclusion

This chapter delved into the Metaverse as an immersive internet iteration that extends beyond traditional online interactions, promising novel, immersive cultural experiences. Although the Metaverse itself remains a work-in-progress vision, the technological advances it spurs are already changing the experiences and expectations of cultural tourists in many ways. This requires cultural institutions like GLAMs to rethink their offerings:

> Museums have little choice but to embrace rapid change in order to maintain strong ties and connections to their audience, actual and potential, real and virtual. As computational culture causes radical change, it is creating a digital ecosystem where art and culture is everywhere and everyone – in every click in multiple forms across the internet and social media. (Bowen & Giannini, 2019: 561)

The typology of immersive technology experiences presented in this chapter suggests that opportunities for enhancing and augmenting cultural tourism experiences abound. These offerings can open cultural tourism to new segments of visitors, including younger, technology-savvy cultural tourists, individuals with disabilities and neurodiverse individuals. This promises that immersive technologies can make cultural tourism more accessible and relevant to a broader group of visitors. Extending GLAM offerings to digital realms can also have a democratising effect as they are often freely accessible and do not require expensive trips. At the same time, the realisation of the 'participatory museum' seems to finally be within reach as Metaverse experiences are more personalised, engaging and open to the input of those who consume them. This has the potential to encourage greater appreciation of the cultural artefacts and heritage displayed. There are no limits to the number of exhibits in the Metaverse and no environmental costs of huge, usually air-conditioned buildings. The Metaverse also provides a unique platform for the preservation of endangered heritage sites and intangible cultural expressions. Digital twins and VR experiences can archive existing cultural objects and experiences and can help with the restoration of physical objects and sites.

While this notion of a more democratic, participatory and immersive future of GLAMs is exciting, digital innovations always have potential drawbacks. Many of the technologies that together provide the foundation of the Metaverse need to be scrutinised in terms of their implications for security and privacy. Further, while Metaverse experiences might be more sustainable than physical cultural tourism, they are not without environmental impacts given the energy consumption of AI and blockchain technology and the growing mountains of e-waste. There are also clear commercial interests in the development of the Metaverse that will shape its ultimate character. In addition, new digital divides can appear among visitors as well as among GLAMs. The struggles of smaller, less well-funded museums to keep up with yet another technological waves are real, and not all potential visitors have access to the technological infrastructure and devices needed to immerse themselves in cultural experiences in the Metaverse. Regardless of these potential roadblocks, the future of cultural tourism will certainly include immersive technologies in one way or another.

Notes

(1) vangoghexpo.com
(2) momentfactory.com
(3) teamlab.art
(4) www.mnhn.fr
(5) schoenbrunnvr.com
(6) cmnh.org/birdly-experience
(7) mitmuseum.mit.edu/collections/collection/art
(8) holocenter.org/
(9) museumofthefuture.ae/
(10) thebroad.org/visit/mirror-rooms
(11) cestavcase.sk/en
(12) https://collection.sciencemuseumgroup.org.uk/objects/co8030220
(13) https://www.museothyssen.org/en/visit/discover-virtual-visit
(14) https://naturalhistory2.si.edu/vt3/NMNH/z_tour-022.html
(15) nhb.gov.sg/acm/acm-online/virtual-tours/russel-wong-in-kyoto
(16) https://www.museothyssen.org/en/thyssenmultimedia/virtual-tours
(17) thekremercollection.com/
(18) https://lhncbc.nlm.nih.gov/LHC-publications/pubs/MobileAppTurningthePages.html
(19) https://www.meta.com/experiences/pcvr/2581900071911230/
(20) https://www.much.go.kr/newsletter/vol.67/en/index.html
(21) https://museumofcryptoart.com/museum/
(22) https://monaverse.com/

References

Ball, M.L. (2022) *The Metaverse: And How It Will Revolutionize Everything*. Liveright.
Bowen, J.P. and Giannini, T. (2019) The digital future for museums. In T. Giannini and J. Bowen (eds) *Museums and Digital Culture* (pp. 551–577). Springer.
Buhalis, D. and Karatay, N. (2022) Mixed Reality (MR) for Generation Z in cultural heritage tourism towards Metaverse. In J.L. Stienmetz, B. Ferrer-Rosell and D. Massimo (eds) *Information and Communication Technologies in Tourism 2022*. ENTER 2022. Springer.
Buhalis, D., Leung, D. and Lin, M. (2023) Metaverse as a disruptive technology revolutionising tourism management and marketing. *Tourism Management* 97, 104724.
Buhalis, D., Lin, M.S. and Leung, D. (2022) Metaverse as a driver for customer experience and value co-creation: Implications for hospitality and tourism management and marketing. *International Journal of Contemporary Hospitality Management* 35 (2), 701–716.
Carayannis, E.G., Bast, G. and Campbell, D.F.J. (2018) Conclusion: The museum of the future and the future of museums. In G. Bast, E.G. Carayannis and D.F.J. Campbell (eds) *The Future of Museums* (pp. 145–148). Springer Nature.
Cho, Y.H., Wang, Y. and Fesenmaier, D.R. (2002) Searching for experiences: The web-based virtual tour in tourism marketing. *Journal of Travel & Tourism Marketing* 12 (4), 1–17.

Cruz Franco, P.A., Rueda Márquez de la Plata, A. and Gómez Bernal, E. (2022) Protocols for the graphic and constructive diffusion of digital twins of the architectural heritage that guarantee universal accessibility through AR and VR. *Applied Sciences* 12 (17), 8785.

Dinhopl, A. and Gretzel, U. (2016) Selfie-taking as touristic looking. *Annals of Tourism Research* 57, 126–139.

Garau, C. (2017) Emerging technologies and cultural tourism: Opportunities for a cultural urban tourism research agenda. In N. Bellini and C. Pasquinelli (eds) *Tourism in the City* (pp. 67–80). Springer.

Giannini T. and Bowen, J.P. (2022) Museums and digital culture: From reality to digitality in the age of COVID-19. *Heritage* 5 (1), 192–214.

Goldsmith, J. (2021) Welcome to Mark Zuckerberg's 'Metaverse' – Next Facebook chapter is an 'embodied internet you are inside of'. Deadline.com. See https://deadline.com/2021/07/mark-zuckerbergs-metaverse-facebooks-augmented-reality-oculus-1234801863/ (accessed July 2021).

Graziano, T. and Privitera, D. (2020) Cultural heritage, tourist attractiveness and augmented reality: Insights from Italy. *Journal of Heritage Tourism* 15 (6), 666–679.

Gretzel, U., Fuchs, M., Baggio, R., Höpken, W., Law, R., Neidhardt, J. ... and Xiang, Z. (2020) e-Tourism beyond COVID-19: A call for transformative research. *Information Technology & Tourism* 22, 187–203.

Gretzel, U. and Jamal, T. (2009) Conceptualizing the creative tourist class: Technology, mobility, and tourism experiences. *Tourism Analysis* 14 (4), 471–481.

Gretzel, U., Sigala, M., Xiang, Z. and Koo, C. (2015) Smart tourism: Foundations and developments. *Electronic markets* 25, 179–188.

Guo, K., Fan, A., Lehto, X. and Day, J. (2023) Immersive digital tourism: The role of multisensory cues in digital museum experiences. *Journal of Hospitality & Tourism Research* 47 (6), 1017–1039.

Guttentag, D. (2022) Virtual reality and the end of tourism? A substitution acceptance model. In Z. Xiang, M. Fuchs, U. Gretzel and W. Höpken (eds) *Handbook of e-Tourism* (pp. 1901–1919). Springer International Publishing.

Hausmann, A. and Schuhbauer, S. (2021) The role of information and communication technologies in cultural tourists' journeys: The case of a World Heritage Site. *Journal of Heritage Tourism* 16 (6), 669–683.

Hurst, W., Spyrou, O., Tekinerdogan, B. and Krampe, C. (2023) Digital art and the Metaverse: Benefits and challenges. *Future Internet* 15 (6), 188.

Huynh-The, T., Gadekallu, T.R., Wang, W., Yenduri, G., Ranaweera, P., Pham, Q.V., ... and Liyanage, M. (2023) Blockchain for the metaverse: A review. *Future Generation Computer Systems* 143, 401–419.

Itani, O.S. and Hollebeek, L.D. (2021) Light at the end of the tunnel: Visitors' virtual reality (versus in-person) attraction site tour-related behavioral intentions during and post-COVID-19. *Tourism Management* 84, 104290.

Jankova, L., Auzina, A. and Zvirbule, A. (2023) Regional smart cultural tourism destinations in a region of Latvia. *Worldwide Hospitality and Tourism Themes* 15 (5), 507–516.

Jenkins, H. (2006) *Fans, Bloggers, and Gamers: Exploring Participatory Culture*. New York University Press.

Johnson, C. (2023) Museum technology trends: A force for social and community impact. See www.electrosonic.com/blog/museum-technology-trends-a-force-for-social-and-community-impact (accessed March 2024).

Jung, Y. (2023) Current use cases, benefits and challenges of NFTs in the museum sector: toward common pool model of NFT sharing for educational purposes. *Museum Management and Curatorship* 38 (4), 451–467.

Kang, D., Choi, H. and Nam, S. (2022) Learning cultural spaces: A collaborative creation of a virtual art museum using roblox. *International Journal of Emerging Technologies in Learning* 17 (22), 232–245. https://doi.org/10.3991/ijet.v17i22.33023

Kang, M. and Gretzel, U. (2012) Effects of podcast tours on tourist experiences in a national park. *Tourism Management* 33 (2), 440–455.

Katsoni, V., Upadhya, A. and Stratigea, A. (2017) *Tourism, Culture and Heritage in a Smart Economy: Third International Conference IACuDiT, Athens 2016*. Springer.

Koo, C., Kwon, J., Chung, N. and Kim, J. (2023) Metaverse tourism: Conceptual framework and research propositions. *Current Issues in Tourism* 26 (20), 3268–3274.

Kozinets, R.V. (2019) Consuming technocultures: An extended JCR curation. *Journal of Consumer Research* 46 (3), 620–627.

Lee, H., Jung, T.H., Dieck, M.C.T. and Chung, N. (2020) Experiencing immersive virtual reality in museums. *Information & Management* 57 (5), 103229.

Lee, H.K., Park, S. and Lee, Y. (2022) A proposal of virtual museum metaverse content for the MZ generation. *Digital Creativity* 33 (2), 79–95.

Liu, X., Wang, D. and Gretzel, U. (2022) On-site decision-making in smartphone-mediated contexts. *Tourism Management* 88, 104424.

Lu, S.E., Moyle, B., Reid, S., Yang, E. and Liu, B. (2023) Technology and museum visitor experiences: A four stage model of evolution. *Information Technology & Tourism* 25, 151–174.

Matteucci, X. and Smith, M.K. (2024) *The Creative Tourist: A Eudaimonic Perspective*. Leeds: Emerald Publishing Limited.

Onix (2024) Virtual Reality (VR) in Museums – The Definitive Guide. See https://onix-systems.com/blog/using-virtual-reality-for-museums (accessed April 2024).

Pine, B.J. and Gilmore, J.H. (1998) Welcome to the experience economy. *Harvard Business Review* 76 (4), 97–105.

Popoli, Z. and Derda, I. (2021) Developing experiences: Creative process behind the design and production of immersive exhibitions. *Museum Management and Curatorship* 36 (4), 384–402.

Ponsignon, F. and Derbaix, M. (2020) The impact of interactive technologies on the social experience: An empirical study in a cultural tourism context. *Tourism Management Perspectives* 35, 100723.

Power, D. and Teigland, R. (2013) Postcards from the Metaverse: An introduction to the immersive internet. In R. Teigland and D. Power (eds) *The Immersive Internet* (pp. 1–12). Palgrave Macmillan.

Pratisto, E.H., Thompson, N. and Potdar, V. (2022) Immersive technologies for tourism: A systematic review. *Information Technology & Tourism* 24 (2), 181–219.

Program-Ace.com (2024) Metaverse vs. Virtual Reality: Differences & Similarities. See https://program-ace.com/blog/metaverse-vs-virtual-reality (accessed March 2024).

Richards, G. (2018) Cultural tourism: A review of recent research and trends. *Journal of Hospitality and Tourism Management* 36, 12–21.

Roque, M.I., Campos, A.C., Almeida, S. and Pasandideh, S. (2024) Transforming museum management through ICT adoption: An analysis of the Portuguese context during the COVID-19 pandemic. *Journal of Heritage Tourism*. https://doi.org/10.1080/1743873X.2024.2331239

Sánchez-Amboage, E., Enrique Membiela-Pollán, M., Martínez-Fernández, V.A. and Molinillo, S. (2023) Tourism marketing in a metaverse context: The new reality of European museums on meta. *Museum Management and Curatorship* 38 (4), 468–489.

Sigala, M. (2005) New media and technologies: Trends and management issues for cultural tourism. In M. Sigala and D. Leslie (eds) *International Cultural Tourism: Management, Implications and Cases* (pp. 167–180). Elsevier Butterworth-Heinemann.

Simone, C., Cerquetti, M. and La Sala, A. (2021) Museums in the infosphere: Reshaping value creation. *Museum Management and Curatorship* 36 (4), 322–341.

Stankov, U., Gretzel, U. and Filimonau, V. (2022) *The Mindful Tourist: The Power of Presence in Tourism*. Emerald Publishing.

Stein, R. (2012) Chiming in on museums and participatory culture. *Curator: The Museum Journal* 55 (2), 215–226.

Sustacha, I., Banos-Pino, J.F. and Del Valle, E. (2023) The role of technology in enhancing the tourism experience in smart destinations: A meta-analysis. *Journal of Destination Marketing & Management* 30, 100817.

Techopedia (2024) Metaverse. See www.techopedia.com/definition/34708/metaverse (accessed April 2024).

TechTarget (2024) What is the Metaverse? An explanation and in-depth guide. See www.techtarget.com/whatis/feature/The-metaverse-explained-Everything-you-need-to-know (accessed March 2024).

Thiel, S. and Bernhardt, J. (2023) *AI in Museums: Reflections, Perspectives and Applications.* Transcript Verlag.

UN Tourism (2018) Technology and innovation seen as a big boost to cultural tourism. See www.unwto.org/global/press-release/2018-10-09/technology-and-innovation-seen-big-boost-cultural-tourism (accessed March 2024).

Werthner, H. (2022) e-Tourism: An informatics perspective. In Z. Xiang, M. Fuchs, U. Gretzel and W. Höpken (eds) *Handbook of e-Tourism* (pp. 3–22). Springer International Publishing.

Wong, A. and Wong, J. (2024) Service robot acceptance in museums: An empirical study using the service robot acceptance model (sRAM). *Journal of Services Marketing.* https://doi.org/10.1108/JSM-05-2023-0183.

Zhou, Y., Chen, J. and Wang, M. (2022) A meta-analytic review on incorporating virtual and augmented reality in museum learning. *Educational Research Review* 36, 100454.

Zhu, H.Y., Hieu, N.Q., Hoang, D.T., Nguyen, D.N. and Lin, C.-T. (2024) A human-centric Metaverse enabled by brain-computer interface: A survey. *IEEE Communications Surveys & Tutorials.* https://doi.org/10.1109/COMST.2024.3387124.

Part 4

General Outlook

12 Quality of Life and Cultural and Heritage Tourism

Muzaffer Uysal and Jiahui Wang

Introduction

Quality of life (QoL) has gained significant attention in academic research and policy discussions in recent decades. One area that has been recognised for its potential impact on QoL is cultural and heritage tourism. Cultural and heritage tourism involves various activities including learning, exploration, experience and consumption of tangible and intangible cultural and heritage resources in different destinations.

Cultural and heritage tourism holds a unique position as it not only offers opportunities for leisure and recreation but also facilitates a deeper understanding and connection to the culture of a specific place. This form of tourism offers visitors opportunities to engage with diverse cultural experiences, such as visiting historical sites, attending traditional festivals or exploring museums and art galleries. These experiences contribute to personal growth, intellectual stimulation and a broader understanding and appreciation of different cultures. Engaging in cultural activities can enhance individuals' QoL by providing a sense of fulfillment, meaning and enrichment. Cultural tourism activities foster a sense of belonging, inclusiveness and social cohesion, which are essential components of QoL. Furthermore, it can contribute to community development, supporting local economies, preserving cultural heritage and enhancing the overall well-being of residents. This broader perspective of cultural tourism, as advocated by Richards (2018), includes not only visiting sites and monuments but also immersing oneself in everyday culture, creativity, and ways of life in each destination. Such experiences can profoundly impact individuals' QoL by enriching their lives through exposure to diverse cultures, fostering personal growth and enhancing overall life satisfaction.

Understanding the relationship between cultural tourism and QoL is crucial for policymakers, and tourism practitioners. Organisations like UNESCO have emphasised the conservation and enhancement of cultural heritage through approaches that promote cultural competence and creativity, thereby improving the QoL and well-being of both residents and visitors inclusively (Milan, 2017). Despite the increasing attention given to

QoL and tourism development, the specific connection between cultural and heritage tourism and QoL has been relatively underexplored in scholarly research. Therefore, it is important to investigate this nexus to obtain a comprehensive understanding of cultural and heritage tourism research. To address these gaps, this chapter combines bibliometric analysis and thematic analysis to identify the academic evolution and noticeable themes in the target field.

We selected the database Web of Science (WoS) to search the related topics. During the search process, the keywords 'cultural tourism' in 'any field' was employed. After literature screening, a total of 1262 articles published during 1985–2023 were maintained for analysis. This chapter used a mixed-methods approach, integrating both bibliometric analysis and thematic analysis, to mitigate the bias originating from a single-method review (Collins & Fauser, 2005). The bibliometric analysis involves the utilisation of statistical and mathematical methodologies to scrutinise the progression of a research domain, delineated through its conceptual, intellectual and social architectures (Zupic & Čater, 2015). This technique also contributes to forecast past, present and future publishing trends. Thematic analysis is employed for analysing qualitative data, identifying the patterns in the data to find themes, and generating a theoretical understanding (Corbin & Strauss, 1990). In this chapter, thematic analysis was applied to identify the noticeable themes and the nexus between QoL and cultural tourism. The results are presented in the following sections: (1) the academic evolution of cultural and heritage tourism; (2) the nexus of QoL and cultural and heritage tourism; and (3) areas of future research and practices for fostering the QoL and sustainability of cultural and heritage tourism. The VOS Viewer software was used to construct visualisation networks.

Academic Evolution of Cultural and Heritage Tourism

In this section, we conducted temporal analysis and thematic analysis to scrutinise the academic temporal evolution and research themes in the cultural and heritage tourism field.

Temporal evolution

Figure 12.1 presents a map of the temporal network of authors' keywords co-occurrence, which conveys the academic evolution history of cultural and heritage tourism. By conducting a co-occurrence analysis of 2847 keywords, it was determined that 123 main author keywords met the established threshold for the minimum number of occurrences (n = 5). To explore how the research topics of interest changed over time from 1985 to 2023, the study period was divided into three distinct sub-periods: 1985–2014, 2015–2017 and 2018–2023. Between 1985 and 2014, the main

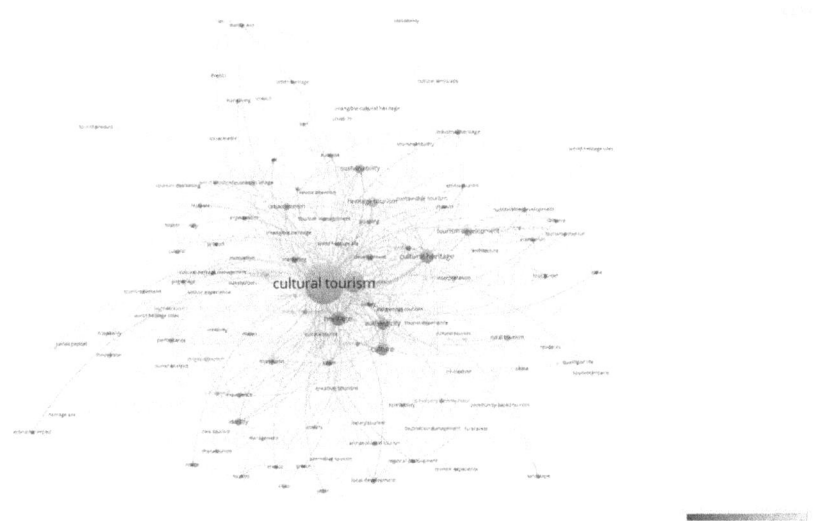

Figure 12.1 Temporal network

keywords such as 'heritage', 'development', 'motivations', 'identity', 'impact', 'marketing' and 'industry' were prominently published. In the subsequent period, from 2015 to 2017, keywords including 'cultural tourism', 'sustainable development', 'indigenous tourism', 'innovation', 'quality of life' and 'museums' gained significance, represented by a greenish colour. Since 2018, the main keywords shown in yellow, have shifted towards topics such as 'creative tourism', 'co-creation', 'visitor experience', 'intangible cultural heritage', 'COVID-19', 'residents' and 'well-being'.

The temporal network analysis of research topics reveals notable insights. First, researchers in the first sub-period primarily approached the topic from a macro level, focusing on areas such as the impact of cultural tourism (Viu *et al.*, 2008), the establishment of development rules for cultural tourism (Liao & Wang, 2011), tourists' motivations (Woosnam *et al.*, 2009) and cultural identities (Stebbins, 1997). These studies aimed to understand the broader implications and overarching dynamics of cultural and heritage tourism during that time frame.

Second, in the sub-period of 2015–2017, researchers exhibited an increased interest in sustainable-related issues within cultural and heritage tourism. They focused on identifying generalised strategies and context-specific measurements for sustainability (Surinach & Wober, 2017). More specifically, there was a growing emphasis on sustaining tourism businesses through innovation (Martinez-Perez *et al.*, 2016) and protecting cultural and heritage resources through tourism development (Whitney-Squire, 2016). Another emerging theme in this sub-period was the examination of QoL aspects within cultural and heritage tourism (Gullion *et al.*, 2015).

Finally, in the most recent sub-period, the research frontiers in cultural and heritage studies have shifted towards the micro level. This is evidenced by the emergence of empirical research that aims to measure and test connections between complex constructs (Xu *et al.*, 2020). For example, there is a growing focus on the topics of co-creation experience and creative tourism (e.g. Richards, 2020; Ross & Saxena, 2019), highlighting a shift towards understanding the active involvement of tourists in shaping their own experiences and promoting creativity in tourism offerings. Furthermore, the target population of research has expanded beyond general visitors and locals. Segmented groups, such as women residents and disabled tourists, are receiving increased attention in recent studies (e.g. Condos & Wirth, 2020; Su *et al.*, 2023). This indicates a greater recognition of the diverse needs, experiences and perspectives of specific demographic groups in the cultural and heritage tourism context.

Noticeable research themes

The review of the results from both bibliometric and thematic analyses suggests that researchers examine cultural and heritage tourism from the following three perspectives: (1) the perspective of tourists, (2) the perspective of locals and (3) the perspective of cultural and heritage tourism product offerings. These perspectives seem to lead to different types of research approaches and methodologies to understand the perspectives. Since these approaches and methodologies are reflective of the perspectives, the content of the study subjects showed a significant amount of variation.

The perspective of tourists

The perspective of tourists as consumers of cultural and heritage resources mostly focused on the behaviour constructs of consumption, such as motivation, perception and attitudinal issues. This perspective represents the demand side of tourism activities.

A significant body of academic research about cultural and heritage tourism motivation is primarily concerned with exploring the dimension of motivation, particularly the typology of tourists as delineated by motivation segmentation. The question of whether tourists of different cultural and heritage sites share similar motivations remains an area of inquiry. Thus, Prada-Trigo *et al.* (2016) segmented heritage city tourists based on three motivation dimensions: cultural, leisure and social and labour issues. Their study revealed that the cultural aspect of motivation holds paramount significance, whereas leisure motivation provides a more comprehensive explanation for tourists' satisfaction. In a subsequent study conducted by Liu *et al.* (2021), a scale of anime tourists' motivation was developed to address the emerging trend of anime tourism, particularly among younger generation visitors. The outcome of their study was a comprehensive five-dimensional motivation scale, encompassing the following dimensions: anime authenticity seeking, anime cultural exploration, novelty, socialisation and escape/relaxation.

Another growing line of research from tourists' perspectives is evolving with perception issues. Within the spectrum of tourists' perceptions, some visitors are specifically driven by the desire to experience the authentic aspects of a destination. The quest for glimpsing the authentic elements can serve as a motivational factor for cultural and heritage tourists. Moreover, the perception of authenticity can significantly influence the overall tourism experience of these visitors. An example of this is demonstrated in the study by Dominguez-Quintero *et al.* (2020), which indicated that both objective authenticity and existential authenticity positively influence the quality of experience and satisfaction among cultural-heritage tourists. Based on their findings, the authors recommended that cultural tourism attractions should strive to offer visitors an authentic and personalised experience to enhance their overall satisfaction.

In line with other areas of tourism research, the research on cultural and heritage tourists also places significant emphasis on variables such as tourists' satisfaction, revisit intention, loyalty and other attitudinal aspects. These variables are widely studied and explored, aiming to understand their antecedents (e.g. Altunel & Erkut, 2015; Dominguez-Quintero *et al.*, 2020).

The perspective of locals

The examination of cultural and heritage tourism from the perspective of locals, representing the supply side of consumption, is concerned with issues such as preservation, protection and impacts of cultural and heritage consumption. Some studies have explored the negative impacts of tourism on cultural resources or stakeholders' lives, including aspects such as change, decline and destruction of cultural value (Frey & Briviba, 2021). Conversely, other studies have identified a positive effect of tourism on cultural conservation, indicating that tourism can contribute to the preservation and protection of culture and heritage. For example, Yun and Zhang (2017) conducted a study involving 290 residents from an ethnic minority area to examine the impact of cultural preservation attitudes. Their findings supported a significant relationship between locals' attitudes toward cultural conservation and the positive perceptions of the tourism impact. Additionally, rather than solely focusing on the perspective of residents, Parga-Dans *et al.* (2020) emphasised the importance of interpreting the value of cultural heritage from different stakeholders' perspectives. They argued that comprehending the diverse interpretations of various stakeholders is crucial for effectively promoting and protecting heritage sites.

The perspective of cultural and heritage tourism product offerings

Understanding and developing cultural and heritage tourism product offerings is another line of research that is essential for the formation of the cultural and heritage tourism experience. The level of development and cultural and heritage product offerings are usually reflective of the locations, where products are staged and consumed. Considering the need for consumption upgrading contemporary studies pay attention to the creative use of cultural and heritage resources (Richards, 2020). Richards (2020) explored the relationship between creativity and tourism, highlighting the importance of integrating creative tourism concepts and the emerging field of creative placemaking. Through this integration, the author identified fundamental design strategies that promote creative development initiatives. Similarly, in a recent study, Richards (2023) indicated that culture is produced not only for tourists but also by and with tourists. This idea somewhat broadens the understanding of what constitutes culture. Another similar study on the topic of co-creation by Ross and Saxena (2019) confirmed that archaeological heritage visitors' participative co-creation contributes to the creation of meaningful experiences.

The Nexus of QoL and Cultural and Heritage Tourism

The preceding sections imply that there exists a natural relationship between types of experiences derived from culture–heritage consumption and the well-being of stakeholders regardless of the study perspective. The nexus of cultural tourism and QoL represents a significant area of study within the field of tourism research. Understanding the relationship between cultural tourism and QoL is important for policymakers, destination managers and researchers seeking to enhance the overall visitor experience and the well-being of both tourists and local communities.

Scholars from diverse fields such as culture, psychology, art and service have conducted a wide range of studies highlighting the significant effect of culture on QoL, for both individuals and groups (e.g. Fujiwara & MacKerron, 2015; Probstl-Haider, 2015). For example, Galloway (2006) conducted a literature review and found research evidence for the impact of cultural participation on individual QoL. Furthermore, Reyes-Martínez *et al.* (2021) utilised Mexico as the case site and confirmed that various forms of cultural participation categories, including book reading, article reading, art class participation and movie watching, are related to individuals' life satisfaction. From the spillover perspective, a study conducted in the UK revealed that various cultural goods can produce positive leisure experiences, thus, further reflecting overall and domain life satisfaction (Wheatley & Bickerton, 2017).

Considering the cultural activities as part of tourism, there exists a natural association between experience and QoL. Cultural and heritage destinations offer opportunities for individuals to access cultural resources. Tourists, driven by their fundamental motivation to experience and consume either the tangible or intangible cultural attractions/products within a destination (Richards, 2018), can benefit from such experiences. In the case of residents living in these destinations, cultural tourism development not only increases their cultural participation but also fosters a sense of identification and pride in the local culture (Bachleitner & Zins, 1999). Therefore, cultural tourism may be linked to the tourists' well-being and residents' QoL. However, it is important to note that empirical evidence supporting this connection remains scarce in existing studies on cultural and heritage tourism (see Table 12.1).

QoL and cultural and heritage tourism from the perspective of tourists

An examination of the limited number of relevant studies reveals that the well-being of cultural tourists can be influenced by experiential factors, including engagement, generativity and place attachment (e.g. Fan & Luo, 2022; Pramanik *et al.*, 2016; Zhang *et al.*, 2023). For instance, Zhang *et al.* (2023) conducted a study involving 429 individuals who had experience in national parks or flower expos and confirmed the effect of cultural ecosystem services on tourists' subjective well-being (SWB). Their findings also tested the predicting role of event attachment and perceived value on SWB. Furthermore, it is worth noting that QoL can also serve as an antecedent in cultural tourism research. A recent study by Peng *et al.* (2023) explored the potential influence of QoL on visitors' revisit intention in two traditional Chinese medicine cultural tourism attractions.

QoL and cultural and heritage tourism from the perspective of locals

Tourism researchers and practitioners have long been interested in understanding the impact of tourism on the various stakeholders in destination areas. Considering the context of tourism communities, the construct of QoL often serves as a consequence of cultural tourism impact or acts as a precursor to attitudinal variables related to tourism development (e.g. Cecil *et al.*, 2010; Yamada *et al.*, 2009). For example, one study examined residents' cultural tourism awareness and its impact on QoL (Cecil *et al.*, 2008). Furthermore, a study conducted by Yamada *et al.* (2009) discussed how cultural tourism impacts residents' QoL. Their findings confirmed a significant relationship among cultural tourism, community pride and community overall QoL. In recent years, scholars focusing on cultural and heritage tourism have endeavored to integrate various variables into a comprehensive model, aiming to

Table 12.1 QoL and cultural and heritage tourism research: Tourist and resident perspectives

Author(s), Date	Purpose	Sample	Findings
QoL and cultural and heritage tourism from the perspective of tourists			
Pramanik et al. (2016)	To test the tourists' nationalism sense (one of the emotional well-being) after visiting museums.	74 new students' parents who visited museums.	After having cultural tourism services, the respondents' sense of nationalism was improved.
Gorchakova and Hyde (2022)	To investigate the impact of cultural exhibition experience on visitor well-being.	27 interviewees participated in semi-structured interviews.	Three dimensions of the cultural exhibition's tourism experience (entertainment, enrichment and emotional engagement) are related to' a person's psychological and emotional well-being.
Fan and Luo (2022)	To establish and validate a research model that examines the relationship between generativity and psychological well-being, while engagement and experience serve as mediators.	17 interviews and 416 questionnaires of art museum tourists.	Generativity directly affects engagement, experience and psychological well-being. Furthermore, engagement and experience serve as mediating variables in the relationship between generativity and psychological well-being.
Peng et al. (2023)	To investigate the mechanisms by which tourists' happiness influences their intention to revisit traditional Chinese medicine and cultural tourism destinations.	311 tourists to China's health resort and medicine museum.	Tourists' happiness contributes to memorable tourism experiences and place attachment, and further evokes revisit intention.
Zhang et al. (2023)	To propose and empirically test a model that explores the relationship between cultural ecosystem service (CES) and visitors' subjective well-being (SWB).	429 visitors to China's National Park and Flower expo.	The perceived value of CES has a strong direct effect on visitors' SWB. In addition, there is a serial mediating effect through place attachment and event attachment.
QoL and cultural and heritage tourism from the perspective of residents			
Cecil et al. (2008)	To evaluate non-economic QOL measures before full implementation of the initiative in 2004 and two years later in 2006.	760 in 2004 and 960 residents in 2006 were collected in Indianapolis.	Residents' understanding of the importance of cultural tourism development positively affects their perception of QoL for two years.
Yamada et al. (2009)	To examine the effect of cultural tourism and four life domains (Community pride, wealth, health perception, safety) on urban residents' life satisfaction.	364 urban residents of a Midwestern city.	The four life domains (community pride, wealth, health perception, safety) and cultural tourism are positively related to urban residents' life satisfaction.

Table 12.1 (Continued)

Author(s), Date	Purpose	Sample	Findings
Shin (2010)	To examine residents' perceptions of cultural tourism.	350 urban residents in Korea.	The mean responses indicate that cultural tourism has improved residents' QoL moderately.
Cecil et al. (2010)	To measure the dimension of QoL, and to evaluate the relationship between the value of cultural tourism and resident's overall QoL.	Residents of Indianapolis participated in a series of surveys conducted from 2004 to 2008.	Cultural tourism has a positive impact on QoL; however, the impact varies randomly over the years, and there is no consistent pattern of either positive or negative effects.
Gullion et al. (2015)	To investigate the relationship between residents' perceived QoL and cultural tourism investment.	Indianapolis QoL survey of 313 residents.	Residents' QoL is positively related to their attitude toward cultural tourism investments.
Tokarchuk et al. (2017)	To study the influence of tourism on residents' well-being in cultural tourism cities.	5226 urban residents in Germany.	All life domains significantly impact individual overall life satisfaction. There exists a positive relationship between tourism intensity and individual life satisfaction among residents in large cities when controlling for other variables.
Li et al. (2021)	To examine the underlying mechanism through which cultural involvement influences residents' attitudes toward tourism development, while spiritual well-being and place attachment serve as mediators.	399 residents in two renowned ancient water villages in China.	The relationship between cultural involvement and attitudes toward tourism development is fully mediated by serial mediation through the constructs of spiritual well-being and place attachment.

Note: The studies included in the table are not intended to be exhaustive in their coverage, but rather serve as illustrative examples.

establish a logical chain that encompasses perception, attitudinal outcomes and the mediating role of QoL. For example, Li *et al.* (2021) have proposed a model that incorporates variables such as engagement, QoL and attitude toward tourism development. Overall, these studies emphasise the importance of understanding residents' perspectives and addressing their concerns to foster sustainable development within cultural and heritage tourism.

Another potential conflict within the cultural tourism community pertains to the issue of over-tourism. Over-tourism is characterised by a volume of tourists that exceeds the carrying capacity of a destination, adversely affecting the QoL for residents. These detrimental effects are particularly pronounced in cultural and heritage sites, where over-tourism poses a significant threat to the preservation of cultural heritage (Adie *et al.*, 2020). Manifestations of these impacts include increased congestion, pollution and crime, as well as escalated living costs, all of which contribute to alterations in the local community's lifestyle. Over-tourism should not be misconstrued as a mere consequence of overcrowding; rather, it is a complex, long-term issue that emerges from the disregard for residents' well-being during tourism development (Kim & Kang, 2020). Mihalic and Kuščer (2021) have identified a dual impact of over-tourism on residents' QoL, whereby it can exert a positive influence through effective destination management, yet also provoke negative repercussions as a result of increased resident irritation.

In summary, empirical evidence has indicated a promising relationship between cultural and heritage tourism and QoL. Nevertheless, there is still significant scope for further exploration and understanding of how cultural tourism and QoL intersect, either from the perspective of tourists or residents. As researchers, we can also delve into the various stakeholders' interpretation of cultural and heritage tourism development simultaneously, thereby gaining deeper insights into the intricate dynamics between cultural tourism and QoL. This knowledge can contribute to the formulation of more effective strategies for sustainable and inclusive tourism development and activities.

Areas of Future Research

Based on the preceding discussion, among others, we focused on four general areas of research interests, namely, creativity and authenticity, well-being, accessibility, and crisis context such as COVID-19 (Table 12.2). Within each area, it is expected that one can identify related specific research topics, reflecting the parameters of this chapter and its focus. We further acknowledge that the scope and direction of future research in cultural and heritage tourism includes a rich portfolio of research areas based on goal, context and target.

Table 12.2 Current research topics and future areas in the studies of cultural tourism and QoL

Current research topics	Emerging themes/trends for future research
The perspective of the tourist: Experiential factors of tourists' QoL – Engagement – Generosity – Place attachment **The perspective of locals:** Antecedents and consequences of locals' QoL – Tourism impacts – Attitudes towards tourism development – Over-tourism and culture preservation	Creativity and authenticity Well-being of diverse stakeholders Accessibility COVID-19 context Cultural and heritage tourism engagement Diversity of product offerings Mindfulness experience in cultural tourism Innovative experience in cultural tourism AI-assisted experience in cultural tourism

Creativity and authenticity

We strongly encourage researchers to further engage in the role and importance of creativity in the consumption of cultural and heritage tourism experiences. Richards and Wilson's (2006) seminal research explored the concept of fostering creativity in visitor experiences as one kind of potential solution to the repetitive replication of culture within the tourism industry. The authors argued that the prevailing tourism model often prioritises standardised and predictable experiences, leading to a lack of innovation and creativity. Consequently, destinations and experiences become increasingly homogenised, resulting in a phenomenon termed 'serial reproduction' of cultural elements. Richards and Wilson (2006) proposed that one way to avoid this commodification and serial reproduction is through the application of creativity, emphasising that tourism innovation should involve consumers actively participating in the creative process itself. This article can be considered early evidence of co-creation within the field of tourism studies. By doing so, we may be able to maintain the authenticity element of cultural and heritage tourism consumption while mitigating the potential over-commodification of these resources through repetitive reproduction. This way of business practice will not only ensure the protection and preservation of these valuable resources but also promote their long-term sustainability.

Moreover, cultural and heritage tourism products can have the unique advantage of offering both creativity and authenticity to visitors inclusively. Although these two elements may appear contradictory, they can coexist and even complement each other within the tourism experience (Daniel, 1996). Creativity often involves innovative and imaginative approaches, while authenticity emphasises the original and genuine aspects of a destination or experience. Both creativity and authenticity are crucial for creating memorable travel experiences. Researchers should address the practical implementation of this idea in future studies, exploring how these two concepts can be effectively combined and leveraged in cultural and heritage tourism.

Indeed, a study conducted by Wang *et al.* (2024) provides an excellent example in this context. Their research explored the paradoxical phenomena of cultural inheritance-based innovation (CIBI), which refers to cultural innovation practices at heritage tourism destinations that are based on cultural inheritance. The study revealed that CIBI involves striking a balance between the paradoxical elements of cultural inheritance and innovation through the effect of an innovation system. Similarly, cultural and heritage tourism academics and industry practitioners should propose specific supportive measures or systems that address the interplay between creativity and authenticity. Understanding and managing this paradoxical interaction between creativity and authenticity is of immense value to destination promoters and practitioners.

Well-being

From the perspective of research themes

Cultural tourism activities possess the potential to enhance the quality of life for all stakeholders and future research might explore this further. The pursuit of authenticity serves as a function to differentiate the demands of cultural tourists into two categories: slow and fast (Calvi *et al.*, 2020). Specifically, cultural tourists with a slow demand orientation exhibit a greater inclination to experience the contemporary life of the destinations they visit. In contrast, those with a fast demand tend to pursue experiences the inauthentic experience in their travel. Future research could investigate the QoL perceptions of these two distinct tourist groups, and provide evidence for cultural tourism planners, enabling them to tailor their offerings to meet the expectations and needs of visitors better, thereby enhancing the overall tourist experience.

In relation to the QoL of destination communities, it is essential for cultural tourism providers to strike a balance between preserving authenticity and ensuring community well-being (Matteucci *et al.*, 2022a). Several cultural tourism destinations have been developed around the traditional lifestyles and cultures of local communities, with these lifestyles serving as pivotal attractions for external visitors. Nonetheless, certain cultural elements may no longer align with contemporary life and societal norms. Consequently, maintaining the purity of authentic traditional culture may inadvertently compromise the QoL of the local inhabitants. Thus, for such cultural destinations, a commitment to sustainable development and preservation is paramount. Drawing from the insights of Matteucci *et al.* (2022a), community-driven slow cultural tourism appears more aligned to bolster community well-being. This development paradigm, underscored by a community-oriented governance approach, prioritises the needs and concerns of local communities over those of tourists and other stakeholders. Furthermore, slow tourism inherently allows locals to engage in lifestyles they find fulfilling.

Beyond the discussion of cultural tourism and QoL from the discrete perspectives of either tourists or residents, it becomes essential for future research to investigate this nexus from an interactive standpoint. Cultural tourism provides opportunities for cultural exchanges between tourists and residents. Upcoming research could explore the impacts of this exchange on both visitors and host communities' QoL. For example, certain forms of cultural tourism, such as creative tourism, slow tourism and proximity tourism, are emerging as resilient forms of cultural tourism capable of fostering greater well-being at the destination level (Matteucci *et al.*, 2022b). Consider creative tourism, which accentuates the active participation of tourists in creative activities and allows visitors to enhance their skills and develop some knowledge about the activity, the local culture and the community (Matteucci, 2018). Similarly, for residents, positive interactions stemming from meaningful participation in the tourism industry can pave the way for direct economic benefits, culminating in tangible economic well-being. Thus, future research in cultural tourism should discern avenues to augment residents' involvement in local cultural tourism activities or to proliferate creative tourism offerings.

While the contributions of cultural tourism to community well-being are widely acknowledged, it is also plausible to consider community well-being as the resource for cultural tourism activities. For instance, the study by Pyke *et al.* (2016) showed the potential of leveraging well-being as a resource for tourism products, the enablers include creating a better destination image and changing consumer climate to healthier lifestyles – such elements drawing visitors to specific cultural and heritage destinations. Admittedly, barriers persist, such as the broad definition of well-being and transient consumer interest in well-being. Nonetheless, Pyke *et al.* (2016) also provided strategies for converting those barriers into enablers to demonstrate the ability to promote well-being as a resource in tourism without inhibitors.

From the perspective of QoL's role

A further examination of the bibliometric analysis reveals that extant studies usually focus on cultural tourism concerning such constructs as satisfaction, place attachment, the nature of experience, and, to a limited extent, the connection between cultural consumption and well-being. However, there is potential to further connect QoL measures with cultural and heritage tourism by drawing upon theories from both the QoL and cultural tourism fields. The bottom-up spillover theory (Neal *et al.*, 1999) and the principles of creative placemaking (Richards, 2020) are two examples of theoretical frameworks that can be employed and extended in this context. For instance, the spillover theory posits that positive experiences in specific life areas, such as cultural life, can positively impact overall life satisfaction (Yamada *et al.*, 2009). Similarly, creative place-making processes involve a relationship among place resources, meanings and creativities, extending beyond the notion of creativity as a mere strategy. This approach has the potential to bring about positive changes and improvements in the QoL of individuals within a cultural place (Richards, 2020).

Additionally, there is limited research examining such constructs with appropriate moderating variables based on target populations, the goals of participants, the sphere of the experience setting and types of cultural and heritage attractions (e.g. museums, historical sites, dark tourism sites, religious sites, music and arts venues, etc.). One possible research direction, as proposed in previous QoL research (Uysal *et al.*, 2016), is to explore the moderating role of cultural background variables, such as trips to destinations that are culturally distal or proximal to the tourists. Conducting research in this direction can help determine whether the relationship between vacation experience and QoL indeed depends on the cultural proximity as perceived by tourists.

Furthermore, within the realm of the connection between the well-being of participants and cultural heritage experience, there is almost no study to examine this connection for disabled consumers (e.g. Condos & Wirth, 2020; Kozlovskaya *et al.*, 2018). The study by Kozlovskaya *et al.* (2018) emphasised the significance of understanding the impact of cultural tourism on personal growth and well-being among disabled individuals. By gaining insights into this relationship, it becomes possible to develop adaptation and rehabilitation programs tailored to the needs of disabled individuals, ultimately improving their overall QoL. There is abundant opportunity to further engage in this line of research.

Accessibility

Both cultural heritage and cultural facilities are vital in supporting the QoL of residents in a community. They provide opportunities for individuals to connect with their cultural roots, express their creativity and engage in cultural activities. Therefore, ensuring equal access to cultural facilities for all individuals is of utmost importance. For example, the Americans with Disabilities Act (ADA, 1990) has established guidelines to ensure that facilities, programs and services provided in a place are accessible to both visitors and employees with special needs. Thus, it is prudent for cultural and heritage facilities to comply with these regulatory guidelines to promote accessibility. Since cultural and heritage sites are often situated in separate locations, destination promoters should integrate cultural and heritage offerings into their design and presentation of resources, facilitating easy navigation and visitation as part of their accessibility efforts. The notion of accessibility should be a shared responsibility among government entities, local authorities and business owners to create easy and broader accessibility and compliance.

Crisis context such as COVID-19

It is essential to recognise that cultural and heritage sites within communities are also vulnerable to crises and economic downturns, natural disasters and pandemics such as COVID-19. These crises can significantly affect cultural and heritage tourism, affecting the economic, social and psychological well-being of tourists and locals at the destination. This vulnerability poses challenges in terms of conservation and maintenance during crises. The closure of cultural and heritage sites and the decline in tourism activities can result in job losses for many individuals, negatively affecting the well-being of those whose livelihoods depend on tourism activities. Furthermore, disasters can lead to changes in tourist preferences and behaviour. Concerns about health and safety may influence the types of cultural and heritage sites that are favoured. Therefore, we need further research that is contingency-based and scenario-oriented to mitigate the possible negative effect of crises on experience formations and the well-being of tourists and communities.

The existing empirical evidence sufficiently supports the assertion that cultural and heritage resources contribute not only to the well-being of visitors but also to the QoL of communities as destinations where such resources are managed and staged. Understanding the nexus of cultural tourism and QoL requires considering both the individual and community perspectives. By recognising the potential positive impacts of cultural tourism on QoL, destination managers and policymakers can develop strategies that maximise the benefits of cultural tourism, enhance visitor experiences and promote the well-being of both tourists and host communities. The future research challenge lies in that we as researchers make sure that the elements of authenticity and creativity are managed, sustained and designed to meet the expectations of current and future visitors while making sure that cultural and heritage tourism product offerings are accessible and in compliance with the rules and regulatory norms of the place as a destination. By doing so, the positive impact of cultural and heritage tourism resources on the enhancement of the well-being of visitors as consumers can be ensured.

References

Adie, B.A., Falk, M. and Savioli, M. (2020) Overtourism as a perceived threat to cultural heritage in Europe. *Current Issues in Tourism* 23 (14), 1737–1741.

Altunel, M.C. and Erkut, B. (2015) Cultural tourism in Istanbul: The mediation effect of tourist experience and satisfaction on the relationship between involvement and recommendation intention. *Journal of Destination Marketing & Management* 4 (4), 213–221.

Bachleitner, R. and Zins, A.H. (1999) Cultural tourism in rural communities: The residents' perspective. *Journal of Business Research* 44 (3), 199–209.

Calvi, L., Moretti, S., Koens, K. and Klijs, J. (2020) The future of cultural tourism: Steps towards resilience and future scenarios. *ENCATC Magazine-The European Network on Cultural Management and Policy* 2020 (2), 34–4.

Cecil, A.K., Fu, Y.Y., Wang, S.S. and Avgoustis, S. (2010) Cultural tourism and quality of life: Results of a longitudinal study. *European Journal of Tourism Research* 3 (1), 54–66.

Cecil, A.K., Fu, Y.Y., Wang, S.S. and Avgoustis, S.H. (2008) Exploring resident awareness of cultural tourism and its impact on quality of life. *European Journal of Tourism Research* 1 (1), 39–52.

Collins, J.A. and Fauser, B.C. (2005) Balancing the strengths of systematic and narrative reviews. *Human Reproduction Update* 11 (2), 103–104.

Condos, B. and Wirth, G. (2020) The role of Nadasdy castle in tourism of Sarvar-The appearance of disabled people in cultural tourism. *Annales-Anali Za Istrske In Mediteranske Studije-Series Historia Et Sociologia* 30 (2), 301–312.

Corbin, J.M. and Strauss, A. (1990) Grounded theory research: Procedures, canons, and evaluative criteria. *Qualitative Sociology* 13 (1), 3–21.

Daniel, Y.P. (1996) Tourism dance performances authenticity and creativity. *Annals of Tourism Research* 23 (4), 780–797.

Dominguez-Quintero, A.M., Gonzalez-Rodriguez, M.R. and Paddison, B. (2020) The mediating role of experience quality on authenticity and satisfaction in the context of cultural-heritage tourism. *Current Issues in Tourism* 23 (2), 248–260.

Fan, Y.L. and Luo, J.M. (2022) Impact of generativity on museum visitors' engagement, experience, and psychological well-being. *Tourism Management Perspectives* 42 (14), Article 100958.

Frey, B.S. and Briviba, A. (2021) Revived Originals – A proposal to deal with cultural overtourism. *Tourism Economics* 27 (6), 1221–1236.

Fujiwara, D. and MacKerron, G. (2015) *Cultural Activities, Artforms and Wellbeing*. Arts Council England.

Galloway, S. (2006) Cultural participation and individual quality of life: A review of research findings. *Applied Research in Quality of Life* 1, 323–342.

Gorchakova, V. and Hyde, K.F. (2022) The impact on well-being of experiences at cultural events. *Event Management* 26 (1), 89–106.

Gullion, C., Hji-Avgoustis, S., Fu, Y.Y. and Lee, S. (2015) Cultural tourism investment and resident quality of life: A case study of Indianapolis, Indiana. *International Journal of Tourism Cities* 1 (3), 184–199.

Kim, S. and Kang, Y. (2020) Why do residents in an overtourism destination develop anti-tourist attitudes? An exploration of residents' experience through the lens of the community-based tourism. *Asia Pacific Journal of Tourism Research* 25 (8), 858–876.

Kozlovskaya, S.N., Anikeeva, O.A., Sizikova, V.V., Shimanovskaya, Y.V. and Maksimova, E.V. (2018) The influence of cultural tourism on the disabled people's personal advancement and evaluation of the quality of life. *Modern Journal of Language Teaching Methods* 8 (12), 342–351.

Li, J.Y., Pan, L. and Hu, Y. (2021) Cultural involvement and attitudes toward tourism: Examining serial mediation effects of residents' spiritual wellbeing and place attachment. *Journal of Destination Marketing & Management* 20 (10), Article 100601.

Liao, D. and Wang, P. (2011) An assessment of sustainability for tourism development-The case of American cultural heritage sites. 5th International Symposium on Green Hospitality and Tourism Management (Guang Zhou, 25–27 June).

Liu, S., Lai, D., Huang, S.S. and Li, Z.Y. (2021) Scale development and validation of anime tourism motivations. *Current Issues in Tourism* 24 (20), 2939–2954.

Martinez-Perez, A., Garcia-Villaverde, P.M. and Elche, D. (2016) The mediating effect of ambidextrous knowledge strategy between social capital and innovation of cultural tourism clusters firms. *International Journal of Contemporary Hospitality Management* 28 (7), 1484–1507.

Matteucci, X. (2018) Flamenco, tourists' experiences, and the meaningful life. In M. Uysal, J. Sirgy and S. Kruger (eds) *Managing Quality of Life in Tourism and Hospitality* (pp. 10–23). CAB International.

Matteucci, X., Koens, K., Calvi, L. and Moretti, S. (2022a) Envisioning the future of cultural tourism. *Futures* 142, Article 103013.

Matteucci, X., Nawijn, J. and Von Zumbusch, J. (2022b) A new materialist governance paradigm for tourism destinations. *Journal of Sustainable Tourism* 30 (1), 169–184.

Mihalic, T. and Kuščer, K. (2021) Can overtourism be managed? Destination management factors affecting residents' irritation and quality of life. *Tourism Review* 77 (1), 16–34.

Milan, S.B. (2017) World heritage management plans as opportunities to foster cultural sustainable development: Fact or myth? 1st International IEREK Conference on Cultural Sustainable Tourism (CST) (Thessaloniki, 27–29 November).

Neal, J.D., Sirgy, M.J. and Uysal, M. (1999) The role of satisfaction with leisure travel/tourism services and experience in satisfaction with leisure life and overall life. *Journal of Business Research* 44 (3), 153–163.

Parga-Dans, E., Gonzalez, P.A. and Enriquez, R.O. (2020) The social value of heritage: Balancing the promotion-preservation relationship in the Altamira world heritage site, Spain. *Journal of Destination Marketing & Management* 18 (13), Article 100499.

Peng, J.M., Yang, X.Y., Fu, S.H. and Huan, T.C. (2023) Exploring the influence of tourists? Happiness on revisit intention in the context of traditional Chinese medicine cultural tourism. *Tourism Management* 94 (17), Article 104647.

Prada-Trigo, J., Perez Galvez, J.C., Lopez-Guzman, T. and Pesantez Loyola, S.E. (2016) Tourism and motivation in cultural destinations: Towards those visitors attracted by intangible heritage. *Almatourism-Journal of Tourism Culture and Territorial Development* 7 (14), 17–37.

Pramanik, P.D., Ingkadijaya, R. and Gantina, D. (2016) Could nationalism sense be reached through cultural tourism activity? 1st International Conference on Tourism Gastronomy and Tourist Destination (ICTGTD) (Djakarta, 14–15 November).

Probstl-Haider, U. (2015) Cultural ecosystem services and their effects on human health and well-being – A cross-disciplinary methodological review. *Journal of Outdoor Recreation and Tourism-Research Planning and Management* 10, 1–13.

Pyke, S., Hartwell, H., Blake, A. and Hemingway, A. (2016) Exploring well-being as a tourism product resource. *Tourism Management* 55, 94–105.

Reyes-Martínez, J., Takeuchi, D., Martínez-Martínez, O.A. and Lombe, M. (2021) The role of cultural participation on subjective well-being in Mexico. *Applied Research in Quality of Life* 16 (3), 1321–1341.

Richards, G. (2018) Cultural tourism: A review of recent research and trends. *Journal of Hospitality and Tourism Management* 36, 12–21.

Richards, G. (2020) Designing creative places: The role of creative tourism. *Annals of Tourism Research* 85 (11), Article 102922.

Richards, G. (2023) Place, culture, and quality of life. In M. Uysal and M.J. Sirgy (eds) *Handbook of Tourism and Quality-of-Life Research II: Enhancing the Lives of Tourist, Residents of Host Communities and Service Providers* (pp. 37–48). Springer.

Richards, G. and Wilson, J. (2006) Developing creativity in tourist experiences: A solution to the serial reproduction of culture? *Tourism Management* 27 (6), 1209–1223.

Ross, D. and Saxena, G. (2019) Participative co-creation of archaeological heritage: Case insights on creative tourism in Alentejo, Portugal. *Annals of Tourism Research* 79 (14), Article 102790.

Shin, Y. (2010) Residents' perceptions of the impact of cultural tourism on urban development: The case of Gwangju, Korea. *Asia Pacific Journal of Tourism Research* 15 (4), 405–416.

Stebbins, R.A. (1997) Identity and cultural tourism. *Annals of Tourism Research* 24 (2), 450–452.

Su, M.M., Wall, G., Ma, J.F., Notarianni, M. and Wang, S.G. (2023) Empowerment of women through cultural tourism: perspectives of Hui minority embroiderers in Ningxia, China. *Journal of Sustainable Tourism* 31 (2), 307–328.

Surinach, J. and Wöber, K. (2017) Introduction to the special focus: Cultural tourism and sustainable urban development. *Tourism Economics* 23 (2), 239–242.

Tokarchuk, O., Gabriele, R. and Maurer, O. (2017) Development of city tourism and well-being of urban residents: A case of German Magic Cities. *Tourism Economics* 23 (2), 343–359.

Uysal, M., Sirgy, M.J., Woo, E. and Kim, H. (2016) Quality of life (QOL) and well-being research in tourism. *Tourism Management* 53, 244–261.

Viu, J.M., Fernandez, J.R. and Caralt, J.S. (2008) The impact of heritage tourism on an urban economy: The case of Granada and the Alhambra. *Tourism Economics* 14 (2), 361–376.

Wang, M.Y., Li, Y.Q., Ruan, W.Q., Zhang, S.N. and Li, R. (2024) Influencing factors and formation process of cultural inheritance-based innovation at heritage tourism destinations. *Tourism Management* 100, Article 104799.

Wheatley, D. and Bickerton, C. (2017) Subjective well-being and engagement in arts, culture and sport. *Journal of Cultural Economics* 41 (1), 23–45.

Whitney-Squire, K. (2016) Sustaining local language relationships through indigenous community-based tourism initiatives. *Journal of Sustainable Tourism* 24 (8–9), 1156–1176.

Woosnam, K.M., McElroy, K.E. and Van Winkle, C.M. (2009) The role of personal values in determining tourist motivations: An application to the Winnipeg fringe theatre festival, a cultural special event. *Journal of Hospitality Marketing & Management* 18 (5), 500–511.

Xu, Y.C., Zhang, H.Z., Lu, L. and Zha, X.L. (2020) Cultural tourists' satisfaction from a leisure experience perspective: An empirical study in China. *Journal of China Tourism Research* 16 (3), 368–390.

Yamada, N., Heo, J., King, C. and Fu, Y.Y. (2009) Life satisfaction of urban residents: Do health perception, wealth, safety, community pride, and cultural tourism matter? International CHRIE Conference-Refereed Track, 24 (Amherst, 29 July–1 August).

Yun, H.J. and Zhang, X. (2017) Cultural conservation and residents' attitudes about ethnic minority tourism. *Tourism and Hospitality Research* 17 (2), 165–175.

Zhang, H.M., Zhang, J.H. and Cai, L.P. (2023) Effects of cultural ecosystem services on visitors' subjective well-being: Evidences from China's national park and Flower Expo. *Journal of Travel Research* 62 (4), 768–781.

Zupic, I. and Čater, T. (2015) Bibliometric methods in management and organization. *Organizational Research Methods* 18 (3), 429–472.

13 Cultural Tourism as Cultural Adaptation: Urban Design Scenarios for the Future

Maurizio Scarciglia

Modern cinema is rich with examples of films that analyse the psychological condition of being a tourist. In the film *Silent City* directed by Threes Anna (2012), Rosa, a Dutch trainee chef, moves to Tokyo for some time. Her dream is to learn the art of cutting fish from master Hon. Soon Rosa must face the harsh reality of cultural differences, finding herself lonely in a very hierarchical society, imbued with very different cultural values. This sabbatical-exotic experience abroad triggers Rosa's deep sense of intra-personal authenticity (Wang, 1999), making her internship a life changing experience. Her experience in Tokyo is a test to her capacity to deal with the unknown, which brings her to get in touch with her intimate self. Rosa's experience represents a perfect example of how important intra-personal authenticity and tourists' emotional states are for long lasting life memories.

At the beginning of my professional career as an urbanist, I mostly considered toured objects' inauthenticity as something deeply negative with respect to the built environment, its historical layering and evolution. Even though I have been in favour of innovations in the fields of architecture and city planning, I have always considered the authenticity of the toured objects to be fundamental for the tourist and genuinely necessary for an immersive, anthropological experience. At a later stage in my career, I found myself repetitively experiencing the feeling of existential authenticity (Wang, 1999) and its actual disconnection from place authenticity. It then became clear to me that the value of the travel experience lies in the emotions felt and which may help the tourist reconnect with her/his inner self. Cultural tourism – or the experience of detachment from everyday life – has ancient routes in sociology studies. Tourism entails the very need of human beings to escape from their daily routine and social structures (Cohen, 1979) to be able to reconnect with their intimate selves. This, of course, has nothing to do with the authenticity of the environment where the tourist experience takes place, nor with a specific historical moment set in time. Wang (1999: 365) describes this point as follows:

> Even if toured objects are totally inauthentic, seeking otherwise is still possible, because tourists can quest for an alternative, namely, existential authenticity to be activated by the tourist experience. In addition to conventional objective and constructive authenticity, an existential version is a justifiable alternative source for authentic experience in tourism.

In other words, tourists do not seek for an authentic environment, rather for experiencing existential authenticity. In the context of the San Fermin fiestas in Pamplona, Spain, Ravenscroft and Matteucci (2003: 11) similarly observed that 'authenticity resides with the intensity of emotion experienced by the individual'. In the last 15 years, I had the professional opportunity to travel to China on a regular basis. During every trip, once my business obligations were fulfilled, I kept one day for myself to randomly wander around new places. Sometimes I ended up finding myself in informal settlements, in very heterogeneous spaces (Edensor, 2006) such as markets with pungent smells, where animals are slaughtered on the street, with trickles of blood slipping on the sidewalks. Yet, my professional duty was just an alibi. The true reason why I chose to work abroad has been to fulfil my deepest need for travelling around the world, my curiosity for discovering different cultures. I always thought that business travel offered unique, immersive opportunities because no filter separated visitors from the everyday life of locals.

Being alone in a context in which we are unknown to people and where we may not be understood, is to me the ultimate way to embody what cultural tourism is about, namely a balance between curiosity and vulnerability, and the pleasure of being fully detached from our home daily life. My view is that too often cultural tourism is misunderstood in its true meaning. True cultural tourism implies a complete acceptance of differences, the risk of discomfort, experiencing extreme foods, dealing with unconventional habits and giving up everyday life facilities. This renouncement is the only way to ensure the preservation of a place and its culture, thus avoiding that local communities adapt to tourists' cultures.

When tourism operators try to reproduce the domestic comfort of a traveller at a destination, they are violating the very meaning of culture, irreversibly compromising its authenticity. To which extent are cultural tourists' experiences truly authentic and to which extent is cultural tourism impacting on local cultures? If one of the objectives of travel is to seek for existential authenticity, it then appears necessary to apply measures that will facilitate experiences of authenticity in tourism and that will circumvent the predicament of mass tourism. Will the future bring significant transformations to urban destination environments?

It is beyond the scope of this chapter to answer such questions; however, in an attempt at reflecting upon them, here I envisage two utopian future scenarios of physical and experiential transformation of cultural tourism destinations. To do so, this chapter speculates two types of urban tourism environment: the resort city and the small provincial town. Before each fictional account of a future cultural tourism scenario is presented, I first take into consideration current trends in spatial transformations. These two scenarios, which are set in the year 2095, distinctively address the relationship between toured objects and existential authenticity. It is up to the reader to wonder whether these uncertain scenarios will translate into utopian or dystopian tourism experiences.

Resort Cities

When observing the evolution of tourism architecture and resort design, it is difficult to avoid the following consideration: if the authenticity of the object is not the precondition for existential authenticity, why do so many tourism locations seek to reproduce architectural archetypes from the past? Planning for hedonism and escapism has prevailed throughout the past centuries, from holiday areas and wellness locations like the thermal baths in ancient Rome, to nowadays tourism capitals. Wellness, culture and entertainment have become hybridised, following societal transformations. Tourism locations have also followed a rather radical transformation into iconography, following the values of hedonic escapism (Holmqvist et al., 2020). The spatial and aesthetic appearance of tourism destinations has developed into an architectural style that accommodates highly commercial leisure programs. One of the first examples of iconic urbanism for leisure and tourism is found in Las Vegas. Las Vegas became a social experiment, which generated heated debates among urban planners and architects about issues related to classical architecture and traditional iconography. In the Las Vegas context, the 'classic touch' has become a strategy, a fetish tool to convey ideals of wealth and comfort to the masses eager to experience luxury. In the same vein, in the *Dubai Experiment*, Katodrytis (2009: 154) suggests that:

> Everyday life is colonized by fantasy, dominated by escapist dreaming. Both the "authentic" architectural icons, and the simulated architectural icons, such as Disneyland or Las Vegas, are inscribed within the same logic of escapist dreaming. Escapism is an ambivalent, even negative word when juxtaposed against realism or authenticity. Yet we are inescapably escapist. We escape from the given into the desirable through theme parks, shopping malls, and suburban developments.

This escapist dreaming phenomenon has led to a proliferation of new leisure destinations, which seek to camouflage a global tourism product with a thin layer of endogenous cultural identity, most of the time misunderstanding it with superficial adaptations.

Organic urban composition

Although many resorts arguably pay particular attention to site conditions and sustainable environmental strategies, the radical makeovers of many coastal spaces have transformed the original relationship between the seaside and inland areas. In landscape design, programs are envisioned through, inspired or even engineered by actual sites. This is in stark contrast to architecture, where project sites are analysed in light of requested functional programs, and design may even develop from those programs (Waldheim, 2015).

Adapting functional programs to the topography and nature in places follows an exoticising process. Opulent interventions tend to radicalise landscape through the extensive use of environmental technology. New, artificial landscape is thus created. Artificial landscapes compromise and impinge upon perceptions of indigenous environments. Such customisation processes have reached an urban or regional scale. When one is browsing the internet in search of a new 'paradise', one can observe that tour operators and real estate companies present newly developed complexes as promises of 'the relaxing experience of the century'. In the USA, Mexico, Egypt, Thailand, Vietnam, United Arab Emirates, Mauritius and Polynesia, to mention only a few, we find resort developments that boast the same key elements. All resorts present a more or less organic composition, where the shapes of nature are stolen and blown out of scale to recreate unique urban ensembles. The use of organic shapes may stem from a desire of mimesis, which is often replaced by a strident clash between nature and architecture. Urban composition, therefore, ends up exacerbating impressions of artificiality in urban development interventions.

Housing complexes follow rigid location rules to meet expectations of privacy and to allow exclusive access to a 'touristified' nature (e.g. a beach, an infinity pool, a panoramic terrace). Lower-budget properties tend to be closer to infrastructural nodes, far from the most exclusive portions of the resort. They include condominium complexes or hotels, which offer fewer facilities, and which lie at a greater distance from resort communal services. Similarly, many resorts enjoy a logical organisation of various public services, which are often clustered together, creating a sort of central citadel. Public services are also nested within denser areas, where hotels are concentrated. This is a common feature of commercial developments, such as shopping malls and theme parks.

Artificial landscape and complex hydraulic transformations

In historical complexes like the Villa from Emperor Hadrian in Tivoli, Italy – his personal hideaway from the burdens of Roman politics – each architectural element contributes to blending newly designed pieces with existing nature. The architectural solutions are rather 'extruded' from the

Figure 13.1 Villa Adriana in Tivoli
Source: the author

landscape conditions, adapting and using them as the foundation of the architectural concept itself (see Figure 13.1).

On the contrary, one of the most emblematic features of contemporary resorts is the obsession with fake natural environments. Nowadays, golf resorts, extensive sports facilities and urban developments with massive residential densities and infrastructure have transformed the way urban planners and architects integrate landscape in their projects, revealing a shift from adapting landscape to imitating it. Newly built resorts echo the functional mechanisms of cities rather than of rural settlements. Many resorts have become water worlds where artificial water systems recreate exotic settings, no matter where the resort is located (e.g. Center Parcs resorts in northern Europe: www.centerparcs.com).

For instance, the Laguna Phuket resort complex in Thailand (www.lagunaphuket.com), which was built on formerly polluted land, came to life through a massive regeneration process of the coastal hydraulic system. The overall intervention was considered a milestone of ecological urbanism, including an impressive network of islands and water enclaves. In December 2004, because of a tsunami that affected the whole Sumatra area, Laguna Phuket was the set of a global tragedy. This tragedy raised important questions regarding the prevention of natural disasters and uncontrolled coastal transformations: in the face of climate change and its associated natural disasters, should such resorts be rethought in terms of values and planning concepts? Shall we challenge nature instead of following its rules? The contemporary Laguna Phuket resort has expanded to become as large as a city. In some instances, an entire city can turn into a resort. Deviating from their original hotel function, resorts often serve as a tool to control guests' activities (e.g. through prescribed all-inclusive packages). No space is left for unplanned, spontaneous activities. The international tourists' need for comfort has transformed resorts into global places, sometimes even erasing their authentic qualities while satisfying international standards. The tourist can now choose between a cottage on a pier located on an atoll, or a seven-star hotel in a skyscraper erected in the middle of the ocean. Casinos, shopping malls, entertainment centres, restaurant chains and indoor skiing complexes have transformed holiday resorts into contemporary dystopian environments.

Following these trends in spatial urban transformation, I present two utopian accounts for cultural tourism destinations.

2095: Reporting from Dubai

December 2095. It is 20 degrees outside. The coast of Zeeland is already full of tourists from Germany, coming to spend their Christmas holidays in the new floating bungalows that cover the whole coastline. I have just booked my next trip. For many years, I have wished to visit the oldest waterfront resort development from the 21st century: the Dubai Palm Jumeirah. Palm Jumeirah became a model for most coastal redevelopment projects that changed the shape of many global waterfront destinations. Because of changes in global climate, it is now possible to spend a beach holiday almost anywhere. I like to combine leisure with culture, therefore I decided to visit Dubai, the place where it all began in the early years of the 21st century.

Arriving to Dubai is surreal. This city is immense. It is impossible to see its limits from above. Only a desert a little more than a century ago, now lines of towers mark the urban skyline with a sea of lights spanning all the way to the ocean. After the oil crisis, Dubai made huge investments to become a global tourism capital and a pivotal industrial and cultural centre between Asia, Europe and Africa. The airport is now the largest of the planet and one can go nowhere without a stopover here. Activities that happened open air in the past are now almost a faint memory. With temperatures reaching 55 degrees, Dubai has become an indoor city. From the airport, people can jump on the hyperloop that brings them downtown in a jiffy. It is impossible to see any landscape, as the whole urban area has been filled with massive logistic developments. Infinite data centres span kilometres-long profiles. Their dull facades are animated by holograms advertising high-technology pharmaceutical products – the rising sector in the region. During a few short minutes of hyperloop ride, one feels like floating in a train that moves outside the expected laws of gravity. Colours change all around; passengers are bombarded with commercial images and sounds. Arriving at destination, a gloomy light is filtered through dark glased photovoltaic roofs, all with the same amber tone. It feels like arriving on Mars, where the desert light desaturates every object. Even sounds are buffered, giving an impression of living a sub-aquatic life.

The atmosphere reminds me of some dystopian scenes from sci-fi films of the last century, mostly representing nightlife, with hundreds of neon lights and steam from the ventilation shafts from Asian restaurants, expelled onto the street, bustling urban streets where it was not possible to see the sky, hidden by infinite skyscrapers and by artificial light pollution. The historical souk of Dubai looks very different. A gigantic transparent dome covers the whole district, making mechanical climate control possible. Individual buildings do not have air-conditioning anymore.

Everywhere, inside and outside buildings, the average temperature is 20 degrees. Lush plants from all over the world grow within public spaces. Birds chirping resonates from the tree crowns, breaking the constant sound of the water flowing through artificial streams. It feels like the city was built inside a botanical garden. No sounds of cars can be heard. Instead, there is the continuous ambient sound of technology: the buzz of delivery drones and the tones of scanned QR-codes of money transactions. Everything happens indoors and the outdoors is devoid of life. Restaurants from every world cuisine animate street life. Two elegant women are sitting at a table, eating locally produced vegan food while sipping Martinis.

Moving through the old town reminds me of the Foro Romano in Rome. The historical buildings are perfectly preserved by climate control devices, rendering the city into a vast indoor archaeological site. 3D printed structures have a single PVC layer, soundproofing windows from the outdoors. There is no need for localised climate systems because the whole city is now climatically controlled. The tour operator sent me all possible activity offers for the week. There is a tour through the largest museum regional park of the planet. Since the last century, the Gulf region started to invest globally to build up the largest art collection in history. The Louvre in Paris and the British Museum in London pale in comparison to the new mega projects developed in Dubai, Doha and Abu Dhabi. Such massive investments have made the park the leading free-trade art economic zone in the world. The hyperloop has a stop at every museum along the line that, passing through Abu Dhabi, Doha and Dammam, connects the UAE to Qatar and Bahrain.

Another package includes traditional tours such as the desert safari – a non-stop drive through sand dunes where it is prohibited to step out of the air-conditioned car, as the outside temperature is no longer suitable for human life. Camel rides are now organised in the newly built indoor desert dome, with fully equipped glamping sites for an all-inclusive overnight stay, evening barbecue and Berber music. Unfortunately, no star-lit sky is included anymore. The best experience in the city, however, remains the water tour: a trip through all places and attractions where water has become the main feature: The Atlantis Aquarium and Underwater Zoo, the electric boat tour through Dubai Marina, the scenic fountains tour and Palm Jumeirah. With the rise of sea level brought by climate change, Palm Jumeirah has become so popular among tourists that every coastal region of the world now boasts having an even larger and better water resort (Figure 13.2).

The water resort in Dubai was the very first of its kind, now a must see for everyone interested in historical sites. It is possible to spend the night in one of the hotels in the underwater Central Business District. Aquatic life was made possible by building underwater, so that buildings can benefit from the cooler conditions of the Persian Gulf waters and avoid the energy

Figure 13.2 Concept image of the Palm Jumeirah global reproduction and the spread of the Dubai resort city model
Source: NAUTA architecture & research

cost of mechanical ventilation. It is now very common to experience sleeping in a room with a full view over an artificial coral reef or dining in one of the many underwater restaurants. Simpson (2016: 36) notes that 'This city is designed for homo ludens and privileges consumption over production, leisure over labour, and gratification over the daily grind'. At the beginning of the year 2000, Dubai was largely criticised for its monumental spatial transformations and for the resulting social inequality that these new developments had produced. Behind glossy luxury developments are the efforts of thousands of migrants from India, Bangladesh and Pakistan, who worked under unfair conditions, were housed in shanty towns outside the exclusive districts, and far from the eyes of affluent tourists. The government, afflicted by the depletion of its oil reserves, had planned to transform the region into a tourism mecca for the world elite. Simpson further remarks that 'tourists are central to the spectacular recent growth of these locales, rather than a by-product of other metropolitan functions' (2016: 30), thus making tourists the protagonists of the whole planning process. The future would bring forth the same development values – an effort reserved to the richer strata of society. Simpson adds:

> Though I have called them tourist utopias, these cities might better be constructed as dystopian for the manner in which vast resources are deployed to benefit a few at the expense of many others, with often devastating social costs and environmental consequences. (2016: 35)

After almost one century, Dubai has become a multi-ethnic city. Many migrants have managed to find a place here, live in dignity, raise children and provide them with a good education in one of the many university campuses. Because of excellent employment conditions, universities were able to recruit many international academics from all over the world. Over time, social disparities have decreased, and the city has turned into a popular tourism and technology mecca, thus consolidating its identity as a mass cultural tourism destination. Many people talk about the 'Dubai dream' in the same fashion as people referred to the American dream of the early 20th century. Besides the happy stories of migrants climbing the social ladder, Dubai is a place where the world's millionaires focus their investments in the booming real estate market. Lines of luxury villas stand empty and inanimate along kilometres of waterfront. These villas are part of the real estate portfolio of wealthy global multinationals, waiting for the next market boom to sell and invest somewhere else. This city has become the image of capitalism on steroids.

In 2095, Dubai is seen as an exemplary destination for its unique (hyperreal) tourism experiences. Yet, the presence of international business chains, the smell of global foods, the sound of languages from all around the world and the mix of cultures and values, once very much rooted in Islamic traditions, make this place no different from Manhattan in New York or Shanghai Pudong. What could be said about authenticity? If the Dubai experience connotes existential authenticity, I reminisce about distinctive, cultural experiences in a world not yet fully globalised: I miss drinking coffee in Taoyuan, or having to drink the only coffee that I could find and which would rather taste like tea; I remember the magic sunlight filtering from the skylights of an hammam in suburban Istanbul or the human smell of a fully packed subway in Beijing, being surrounded by a homogeneous river of dark haired heads rushing to work. My experiences of existential authenticity were triggered by the contrasting authenticity of the toured objects. Dubai, as a globalised city, I argue, inhibits experiences of existential authenticity.

Small Provincial Towns

Since the beginning of the 21st century, climate disorders have intensified, which has led to serious actions to 'rescue' the planet. Discussions have become more frequent on how to shift from linear economic growth to circular systems, hoping to curtail natural resources exploitation and over consumption. The 20th century saw the emergence of small towns, especially in the tourism market. In the United States, the National Trust Main Street Centre, operated by the National Trust for Historic Preservation, has been fostering the development of small towns' main streets, offering advice on historic preservation as well as on business development (e.g. through the provision of official certification for towns becoming a Main Street community) (Knox & Mayer, 2009). In Italy, the 'Alberghi Diffusi' association was created. This hospitality model was developed by tourism marketing professor Giancarlo Dall'Ara and formally acknowledged in Sardinia with a specific regulation in 1998 (www.alberghidiffusi.it).

Over time, small provincial towns have been suffering from monofunctional planning, economic monopolies, and in turn from unstable economies highly dependent on seasonal tourism activities. Many towns did not survive such pressure and turned into ghost towns. Local governments, compelled to sustain the local economy, have gradually allowed the spread of franchised chains. These commercial chains have served as instruments of economic displacement, akin to invasive species: voracious, indiscriminate and often antisocial (Knox & Mayer, 2009). These commercial formulas have eroded small businesses while narrowing consumer choice. Even today, we find the same Starbucks or McDonald's outlets in most of the small towns along major tourist routes. These outlets work as extractive economic networks, diverting profits away from local communities. At the same time, such outlets undermine authenticity of local products, rendering consumer choice limited and standardised.

Within this globalised context, studies propose new alternative economic models based on social needs rather than economic growth. Many concepts have emerged both within and outside academic circles. For instance, *doughnut economy* approaches (Raworth, 2017) and *degrowth* (Kallis *et al.*, 2014) offer alternative values and practices to foster societal wellbeing. In the same vein, Ray Oldenburg (1999: 16) has explored the idea of 'third place' as the informal places that 'host the regular, voluntary, informal and happily anticipated gatherings of individuals beyond the realm of home and work'. These include, among others, German beer gardens, English pubs, French cafés, Italian bars, coffee houses and bookshops (Knox & Mayer, 2009). Creating third places, as social experiments, aims at transforming contexts with their existing heritage potential, into self-supporting communities.

Alongside the massive urbanisation process of the last century, a new phenomenon of regional network cities has been gaining traction. The idea of connecting provincial towns within a regional network has been valuable, especially in developed rural areas, which have experienced depopulation. Such intra-regional networks have been possible thanks to new mobility systems such as '15-minute cities' (Allam *et al.* 2022), which connect small towns by fast public transport, thus facilitating complementary economies and services. The great leap towards social communities could bring provincial town centres back to life.

2095: Reporting from Rovereto

To envisage a new development paradigm for small towns that strikes a balance between identity preservation and growth, I will guide readers through Rovereto in the Italian province of Trento (region of Trentino Alto Adige) (Figure 13.3). This provincial town, which counted 40,000 inhabitants during the last century, now plays an important role within the cross-border metropolitan region of the provinces of Bolzano, Brescia,

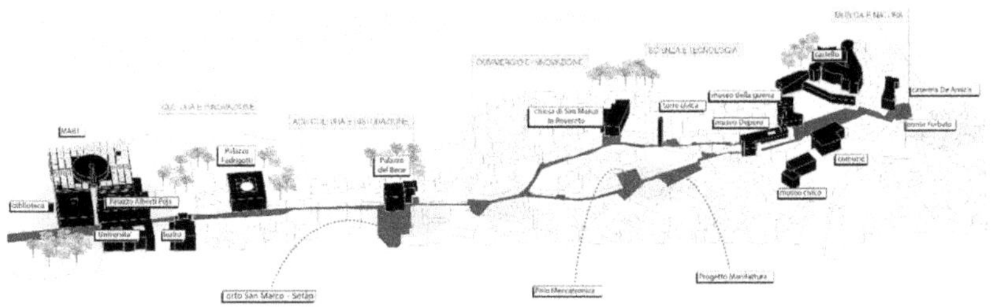

Figure 13.3 Concept image of urban renovation interventions in the historical centre of Rovereto
Source: NAUTA architecture & research

Verona, Innsbruck in Austria, and all the way to Liechtenstein. Despite global warming, the region remains an important destination for winter sports and to mountain lovers. Popular destinations like St. Moritz, St. Anton, Mayrhofen, Merano and Cortina d'Ampezzo are now part of an extensive ski-resorts network, while major cultural and technological investments in the south of the region have made the province of Trento a hub for cultural tourism and technological innovation.

I reach Rovereto by fast train connecting Milan with Bergamo, Rovereto, Trento, Bolzano and Innsbruck. The station lies between the historical town centre and the industrial district. The technology sector around the University of Trento has grown enormously, making the whole region a global leader in robotic technologies and artificial intelligence. This growth has compelled the region to embrace renewable energies, which have now become a symbol of Rovereto. Arriving at the station is a journey back to the Middle Ages. The historical centre of Rovereto is immaculate, perfectly preserved in its structure. No motor vehicles circulate here. The surreal historical gaze at old horse carriages transporting tourists is contrasted by images of electric scooters and self-driving hydrogen-based mini-busses. A student who speaks perfect English, approaches me with a small device. By scanning this device with my smartphone, I am granted access to an online platform where I can find information about tours, activities, venues and restaurants. This app offers a tailored digital tour based on my personal needs and interests. Also, one is given access to an urban game, a bonus system through which anything done in the city translates into 'coins' for discounts, free entrance to attractions and free meals within a network that promotes circularity and social inclusion (Figure 13.4). Everything is locally produced here.

The urban experience is a thick network of spaces that meander between retail, restaurants, clubs, co-working spaces and technology labs run by the University of Trento, which is one of the main investors in the area. The technology sector has become an asset not only for boosting the economy but also as a tool for redefining the town's identity between history and innovation. The urban environment reminds me of the

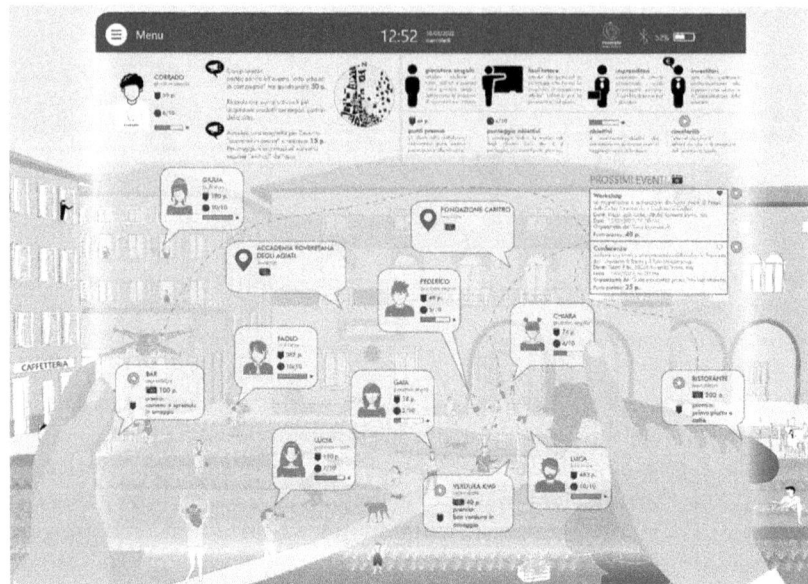

Figure 13.4 Concept image of the interface of 'Stay in the Doughnut': An urban game to stimulate citizens' participation in curating the cultural program of Rovereto
Source: NAUTA architecture & research

bustling streets of Hong Kong, with a 24/7 active plinth and a sequence of images of different spaces and functions. The ground floor of the buildings is filled with public activities, while the historic town centre, in its original form, 'floats' on top. Because its architectural heritage has been well-preserved, one can still experience the town's historic charm. Yet, the presence of technology is felt immediately. During the past decades, the network of local factories, research centres and universities has contributed to building a competitive technology hub within the global market. This mechatronic and robotic focus has grown to the point that the region now competes with traditional research centres in South Korea and Japan. In 2095, in the town centre, it is common to meet Asian engineering students attached to departments of Trento University; these students populate the existing building plinth and make the whole city pulse with creative energy. The same students find lodging opportunities in renovated buildings in the town centre, infusing the indigenous population with a new international and global social layer.

At the beginning of this new social process, changes were highly debated and contested. Many residents considered this social development invasive. They expressed concerns with processes of gentrification, and with the threat of substituting the local community with a new multicultural society. Today, both social groups have merged into a healthy internationalised community. While I was sitting at the terrace of a restaurant

Cultural Tourism as Cultural Adaptation: Urban Design Scenarios for the Future 219

Figure 13.5 Concept image of Piazza Rosmini: The countryside in the city (automated agricultural process)
Source: NAUTA architecture & research

in Piazza Rosmini, I talked to a couple seated at a table nearby. They had a baby in a stroller. Silvia is a native from Rovereto where she grew up. Kim comes from Busan, Korea. He came here as an exchange student. They met in Rovereto and fell in love. He is now an Associate Professor at the University of Trento, responsible for a collaborative project between Italy and Korea on robotic technology applied to neurological surgery.

Piazza Rosmini is open to the natural landscape surrounding the town, making it a link between the town centre and the stunning surrounding nature. It was an empty space for many years, with vacant buildings and no planned activities for locals or tourists. A community deliberation process, supported by key stakeholders and local farmers, reinvented this space, making it an important agricultural hub. Automated production, the use of watering drones, and the work of a local association of inhabitants that cultivate the land made the whole district self-sustainable (Figure 13.5). The revitalisation of Piazza Rosmini facilitated the rebirth of the local cuisine and promoted a new trend of oeno-gastronomical tourism. Several new restaurants surfaced. Every year, the Rovereto Food Fest brings thousands of tourists to town and Rovereto now prides itself on having three Michelin-star restaurants (Figure 13.6).

Silvia and Kim suggested that I skipped coffee at the restaurant and instead moved to Piazza delle Oche. That square is known for Bontadi, the oldest Italian coffee roasting café, which has been famous for its

Figure 13.6 Concept image of Piazza Rosmini: Creation of a catering hub for the Trentino cuisine festival
Source: NAUTA architecture & research

Coffee Museum since 1790. Nowadays, the square hosts an installation of hanging punching bags, filled with recycled coffee powder used to grow mushrooms. Local farmers and Bontadi worked together to establish a local zero-waste farm – a must-see for anybody coming to town. Visitors can harvest their own mushrooms on the square and pay at the café counter (Figure 13.7).

Rovereto has been known for its tradition in art, for being the birthplace of the artist Depero, for its pivotal centre for Futurism, and for hosting Italian literary and other artistic movements since the beginning of the 20th century. In 2002, the city inaugurated the Museum of Modern and Contemporary Art of Trento and Rovereto (MART). Since its opening, a team of curators has transformed the museum into an important experimental centre. The central open hall is now an active public space curated to host astonishing cultural events such as exhibits from renowned international museums, reproductions of real artwork in 3D scale using holograms (Figure 13.8), and installations inspired by the work of Olafur Eliasson (e.g. a botanical garden with a central rainwater collection basin, which offers an oasis of biodiversity in the heart of the city; see Figure 13.9). The museum, which attracts many cultural tourists and art lovers, is now recognised as a leading player among global art institutions.

This urban scenario offers a seamless combination of flashbacks to the past and sudden jumps into an accelerated modernity. Piazza delle Erbe is

Figure 13.7 Concept image of Piazza delle Oche: Mushrooms farming with recycled coffee powder
Source: NAUTA architecture & research

Figure 13.8 Concept image of MART Museum: Hologram reconstruction of artwork from international collections
Source: NAUTA architecture & research

Figure 13.9 Concept image of MART Museum: The museum returned to nature – Landscaping installation by Olafur Eliasson
Source: NAUTA architecture & research

still used for summer screenings by RAM, a festival focused on cinema and archaeology. Here, projections of Federico Fellini's 'La Dolce Vita' at the open-air cinema (Figure 13.10) are followed by a medical congress on robotics, with synchronous presentations from physicians from Tokyo, Rome and Dubai (Figure 13.11).

On the edge of the historic centre, a public road ends at the river Leno. This place has undergone significant changes due to climate change. During the last century, the area suffered from severe drought and the old water line receded significantly, transforming it into a barren landscape. Massive water management efforts, supported by the central government, succeeded in partially restoring the river flow. Nowadays, it is a seasonal waterline, famous for being dry with periodic flooding. During the dry season, the riverbed is transformed into an extensive public space, home of the Rovereto Music Fest (Figure 13.12). During this event, a stage is set at the centre of the riverbed, and while most concert attendees populate the riverbanks, some stand on the balconies of the surrounding buildings. The area is surrounded by historic barracks transformed into hotels, where visitors may experience the atmosphere of former local military life. Furthermore, the lush orchard surrounding the riverbanks brings back to life the old standing tradition of apple harvesting of the region (Figure 13.13).

Rich of its European cultural heritage, the small provincial town of Rovereto has managed to reinvent itself to meet the challenges of the

Cultural Tourism as Cultural Adaptation: Urban Design Scenarios for the Future 223

Figure 13.10 Concept image of Piazza delle Erbe: Open Air Cinema (RAM Festival of Cinema and Archaeology)
Source: NAUTA architecture & research

Figure 13.11 Concept image of Piazza delle Erbe: International Congress on Robotic Surgery
Source: NAUTA architecture & research

Figure 13.12 Concept image of the Leno stream: Rovereto Music Festival on the stream banks
Source: NAUTA architecture & research

Figure 13.13 Concept image of the Leno stream: A resilient landscape integrating historical apple cultivation from Trentino Alto Adige
Source: NAUTA architecture & research

(post)modern, global world. This transformation was thought to be the only way for the town to resist issues of seasonality in tourism. The successful preservation of the local historical heritage, and the mix of tourism activities and technological innovations offer unique experiences of an authentic culture and existential authenticity. This utopian vision of Rovereto represents an attempt at combining two extreme phenomena within cultural tourism, namely the preservation of local identity and globalisation processes. In other words, this utopian vision seeks to accommodate societal changes and technical advancements without undermining historical heritage.

Conclusions

In the film *Lost in Translation* directed by Sofia Coppola (2002), Bob, a jaded movie star, travels to Tokyo to promote Suntory whiskey. Bob is married to Lydia. They have two children and are entering the oblivion of a long-term relationship, devoid of excitement and complicity. Charlotte is a young woman in her twenties, a recent Yale philosophy graduate. Still unemployed and unaware of what she wants in life, she follows her husband John, a photographer sent to Japan for a mission but too busy to share time with her. Bob and Charlotte are both feeling lost in different periods of their lives. They meet in the Hyatt Hotel of Tokyo (ever since one of the most renowned tourism destinations in Tokyo) and share an intimate friendship. The choice of the hotel setting is not coincidental: like airports, hotels are non-places (Augé, 1995) where passers-by cross ways without barely interacting and remaining anonymous. Both characters meet in a country where their native language is hardly understood, and where their lifestyle, culture, food and social habits are at odds with the Japanese culture.

In the Japanese cultural context, Bob and Charlotte get in touch with their authentic selves, an experience that shakes their inner stillness. They manage to see, yet again, their own life from the outside. The tourist experience of existential authenticity seems to be facilitated by a sort of cultural shock, a clash between two cultures where experiences of authenticity play a central role. The experience of toured-object authenticity is what makes tourists different from locals. For instance, if a local experienced the same sense of detachment from the context as a tourist did, he/she would miss the very sense of belonging to his/her sociocultural milieu. Delineating the boundaries between utopian and dystopian visions for cultural tourism is a difficult exercise because both visions are closely intertwined. Because the societal uncertainties that humanity is facing are both elusive and interconnected, one may only imagine the future by simultaneously juxtaposing positive and negative likely transformations.

Technological advancement may be beneficial (e.g. reduction of carbon emissions from traditional mobility systems); yet, we wonder whether such technologies would impinge upon our privacy and freedom. Let's imagine, for example, that we were able to commute by flying around the city. What would be the consequences of having drones at our window? The two futuristic accounts presented here – Dubai as a resort city and Rovereto as a small provincial town, have one thing in common, namely the inescapable force of globalisation and the consequential impossibility of preserving place authenticity in its most traditional sense. In a constantly changing global world, meanings and understandings of authenticity will continue to evolve into various hybrid forms. Paradoxically, while existential authenticity has become a desirable concept underpinning cultural tourism, the cultural homogenising force of globalisation threatens experiences of existential authenticity.

References

Allam, Z., Bibri, S.E., Chabaud, D. and Moreno, C. (2022) The '15-Minute City' concept can shape a net-zero urban future. *Humanities and Social Sciences Communications* 9, 126.
Anna, T. (Director) (2012) *Silent City* [Film]. Pathé Thuis.
Augé, M. (1995) *Non-places: Introduction to an Anthropology of Supermodernity* (Trans. J. Howe). Verso.
Cohen, E. (1979) A phenomenology of tourist experiences. *Journal of the British Sociological Association* 13 (2), 179–201.
Coppola, S. (Director) (2002) *Lost in Translation* [Film]. American Zoetrope/ Elemental Films.
Edensor, T. (2006) Sensing tourist spaces. In C. Minca and T. Oakes (eds) *Travels in Paradox: Remapping Tourism* (pp. 23–46). Rowman and Littlefield.
Holmqvist, J., Ruiz, C.D. and Peñaloza, L. (2020) Moments of luxury: Hedonistic escapism as a luxury experience. *Journal of Business Research* 116, 503–513.
Kallis, G., Demaria, F. and D'Alisa, G. (2014) Introduction: Degrowth. In G. D'Alisa, F. Semaria and G. Kallis (eds) *Degrowth: A Vocabulary for a New Era* (pp. 1–18). Routledge.
Katodrytis, G. (2009) The Dubai experiment. In E. Blum and P. Neitzke (eds) *Dubai: Stadt aus dem Nichts* (pp. 150–156). Birkhäuser.
Knox, P.L. and Myer, H. (2009) *Small Town Sustainability. Economic, Social and Environmental Innovation*. Birkhäuser.
Oldenburg, R. (1999) *The Great Good Place: Cafes. Coffee Shops, Bookstores, Bars, Hair Salons and Other Hangouts at the Heart of a Community*. Marlowe and Company.
Ravenscroft, N. and Matteucci, X. (2003) The festival as carnivalesque: Social governance and control at Pamplona's San Fermin fiesta. *Tourism, Culture & Communication* 4 (1), 1–15.
Raworth, K. (2017) *Doughnut Economics: Seven Ways to Think Like a 21st-Century Economist*. Chelsea Green Publishing.
Simpson, T. (2016) Tourist utopias: Biopolitics and the genealogy of the post-world tourist city. *Current Issues in Tourism* 19 (1), 27–59.
Waldheim, C. (2015) Is landscape urbanism? In G. Doherty and C. Waldheim (eds) *Is Landscape...? Essays on the Identity of Landscape* (pp. 162–189). Routledge.
Wang, N. (1999) Rethinking authenticity in tourism experience. *Annals of Tourism Research* 26 (2), 349–370.

14 Uncharted Territories in Cultural Tourism: Synthesis and Some Personal Reflections

Simone Moretti and Xavier Matteucci

The concept of cultural tourism lies at the intersection of cultural heritage and tourism. It is a fluid concept whose meaning has changed over time and will continue to change as our world evolves. In the past, cultural tourism was associated with the 'high quality' consumption patterns of a rich and educated minority of city dwellers primarily interested in built heritage. Gradually, cultural tourism has shifted to include popular culture (e.g. films, street art, comics, popular music) and residents' everyday spaces and activities (Richards, 2022). Cultural tourism has, therefore, evolved along a reframing of the concept of *culture*, a concept now vaguer and more porous than ever before. In the same vein, cultural heritage, as an essential element of cultural tourism, has also evolved over time and its interpretation has become less authoritative. For instance, cultural heritage has been understood as a process, or a set of practices, revealing different modes of experience and engagement with cultural and natural resources. Cultural heritage has also been considered as a product (e.g. cultural villages, museums, World Heritage sites), which can be developed, controlled and managed. However, a more fluid understanding of cultural heritage as performance makes us question whether these two dimensions can be disassociated. Furthermore, as Ashworth (2015: 177) notes, heritage has 'usually no single clearly recognized producer, consumer or purpose'. The same observation about the blurring distinction between producers and consumers is made by Richards (2021) in the context of cultural tourism. In line with recent conceptualisations of cultural heritage as a democratic process (Smith, 2015), the European Council's Faro Convention (2005: 2) provides a holistic definition of *cultural heritage* as:

> a group of resources inherited from the past which people identify, independently of ownership, as a reflection and expression of their constantly

evolving values, beliefs, knowledge and traditions. It includes all aspects of the environment resulting from the interaction between people and places through time.

Since cultural heritage is a fluid process, we wonder what the role and significance of cultural heritage will be in the future? We also wonder whether extant tangible and intangible heritage will be preserved for generations to come? Another question is what new cultural forms will become heritage and how? These questions are relevant because, as Ashworth (2015: 177) remarks, 'heritage is not socially and politically neutral'. Richards (2021: 1) circumstantiates this view when he writes that countries have 'used culture to forge new identities and create homogeneous national cultures.' When we invited authors to contribute to this volume on the future of cultural tourism, we did not prescribe our own view of what *cultural tourism* is, therefore, the authors of the chapters presented here have looked at cultural tourism from various angles. Furthermore, when we invited authors to contribute to this volume, we did not set a specific time frame of what we understood as 'future'. Near future, mid-range future or far future? Interpretations of the future were, therefore, free and open. Consequently, in this volume, most authors have chosen to explore various cultural tourism topics from a near future perspective.

In the introductory chapter of this book, Matteucci referred to three futuristic visions of cultural tourism, namely a utopian, a dystopian and a heterotopian (see also Matteucci *et al.*, 2022a), which have been useful to situate the work presented in this volume. However, the insights presented in this book point to a complex and intricate network of variegated perspectives, in which both utopian and dystopian visions are enmeshed and act upon each other. This variety of perspectives reveals how challenging it is to imagine the futures of cultural heritage consumption without reflecting on global politics and on the evolution of various modes of tourism governance within their own idiosyncratic contexts. Therefore, it would be simplistic to label the future of cultural tourism merely as either utopian or dystopian. Failing to acknowledge the unpredictability that is intrinsic to any complex sociomaterial network would only produce naive assumptions about the future. If optimism pervades most of the chapters in this volume, a nuanced vision of the futures of cultural tourism should be warranted. In fact, many contributing authors do recognise the interplay of various sociopolitical forces; some are already observable today (e.g. detrimental business and governance practices), and others are yet to materialise (e.g. revitalisation of local heritage).

If the future of cultural tourism won't be fully rosy or fully terrible, then it surely will be something in between. One in-between scenario may consist of a rather bleak future in which anthropocentric disruptions keep affecting eco-systems and livelihoods; however, while disruptions are slow, gradual and uneven, many hopeful, resilient communities are working to reverse the tide. Such communities are found everywhere, and their size and number are likely to grow. Because these communities are and will arguably remain marginal within a global neoliberal political order, in Matteucci et al. (2022a: 6), we used the term 'pockets of resistance' to describe their alternative modes of consumption and governance. We presented these pockets of resistance as spaces imbued with ethical cultural values such as those of solidarity, conviviality, respect for diversity, social justice and collective wellbeing. These pockets of resistance or heterotopias (Foucault, 1986) are spaces of collaboration, creativity and experimentation towards alternative ways of being (or becoming) and doing. Pockets of resistance are akin to municipalist alternative spaces of economic activity imbued with values of democracy and solidarity. Many authors in this edited volume have similarly identified specific cultural values, which constitute defensive mechanisms for local communities to act against the market failures produced by neoliberal policies. Below, we refer to these pockets of resistance as *bubbles of ethical consumption/governance*. If heterotopias are enactments of utopian ideals, heterotopias function as counter-powers to hegemonic discourses and structures, hence as an endeavour to reclaim the rights of communities to democratic and regenerative futures.

Across the many chapters presented in this book, digital technologies have emerged as a powerful and unpredictable force, which is likely to disrupt the way tourists and locals experience, manage, understand and engage with cultural heritage. The crucial role of digital technologies is represented in Figure 14.1 below as either a positive force, a negative force,

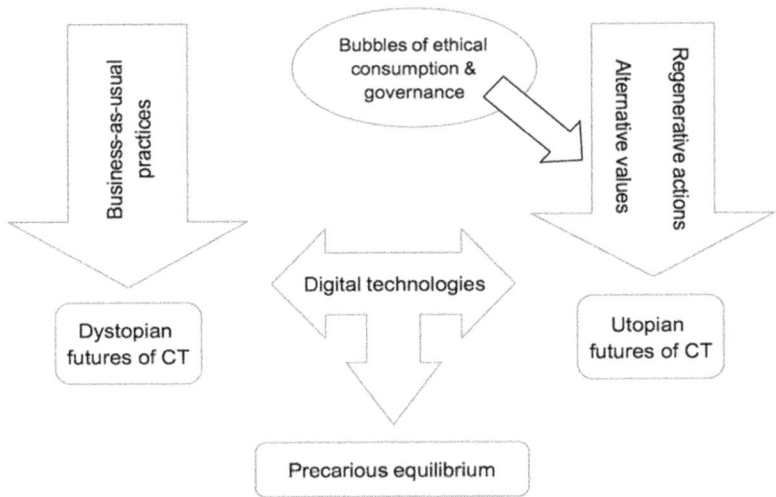

Figure 14.1 Model of possible cultural tourism futures

or both. In the subsequent sections, we discuss the key lessons learnt from the chapters presented in this volume, and conclude with some final thoughts about the future. Table 14.1 provides a synthesis of the key ideas and concepts derived from the chapters.

The Slippery Slope Towards Dystopian Realities

While most contributing authors have looked at the future of cultural tourism through an optimistic lens, many warn that dystopian manifestations are already visible within current social realities and forms of cultural tourism. Dystopian manifestations can be linked to forms of governance that are not geared to fostering community wellbeing and to safeguarding 'authentic' cultural heritage. Some governance issues are illustrated in the Croatian context in Chapter 4. There, Petrić, Mandić and Mikulić observe how the lack of specific protective measures, inadequate community participation, and limited spatial planning have failed to safeguard endogenous cultural heritage in the Dalmatian hinterland. They express concerns over future uncontrolled tourism development, which would result in urban sprawling and disorganised growth patterns. In addition, they contend that prioritising the interests of a few private investors would undermine the local communities' efforts to preserve their cultural heritage. In Chapter 2, Wanner, Shearer Demir and Volić are similarly concerned with top-down governance practices. These authors advocate for a return to more sincere forms of hospitality, which would not only transcend mere commercial transactions between hosts and visitors, but which would also circumvent the ubiquitous trap of cultural commodification. Their concerns echo long-standing debates in academic circles about the detrimental effects of mass cultural tourism on places, with commentators blaming tourists for the commodification and homogenisation of local cultures (e.g. Ashworth, 2015; Kirshenblatt-Gimblett, 1998). Against the dominant neoliberal values underpinning destination governance, Wanner *et al.* argue for establishing networks that support fair profit-sharing and empower local community members. Concerns with poor cultural tourism governance are shared by Timothy, in Chapter 7, who notes the mounting dissatisfaction of residents with the sociocultural and ecological impacts of mass cultural tourism (or overtourism). Timothy questions the use of ill-suited quantitative indicators, which have been traditionally employed by business analysts to measure the success of tourism destinations. In Chapter 6, Farkić connects the influence of capitalism and profit-centred development with the consequences of industrial expansion, urbanisation, and the commodification of global cultural landscapes. Like Wanner and his colleagues, she calls for alternative modes of being and doing within community-based cultural tourism development.

Cultural commoditisation, homogenisation of cultures, appropriation and destruction of heritage (e.g. religious sites), resident alienation and the marginalisation of indigenous groups are among the many issues that tourism is often blamed for, and which represent significant challenges for destination organisations. However, local communities are becoming more protective of their cultural heritage (Raj & Griffin, 2017) and emancipatory movements underpinned by alternative cultural values are gaining momentum worldwide.

Alternative Cultural Values

Cognisant of the risks associated with business-as-usual thinking, and of the imperatives of a regime change, some contributing authors are calling for alternative cultural values to guide cultural heritage and tourism consumption and governance. This ontological concern is reflected beyond the field of cultural tourism with a growing number of scholars urging policymakers and academics to embrace alternative philosophies and practices. For instance, with *hopeful tourism* Pritchard *et al*. (2011) espouse co-transformative learning to create a more just and sustainable tourism future. Both Matteucci *et al*. (2022b) and Guia (2021) suggest that the post-anthropocentric values associated with new materialism and post-humanism should infuse destination governance. New materialist researchers, Matteucci and Smith (2024) assert, do not cast aside the human subject from their analysis; however, they acknowledge that 'human experience is always embodied and entangled with non-human materialities' (2024: 9). In new materialist philosophy, non-human subjects, such as other species and other materialities, are regarded as equally agentic and significant as human beings. The intrinsic value of post-humanist thinking to understanding and managing heritage is also recognised by Sterling (2020). Post-humanism connects closely to the concept of regenerative tourism (Dredge, 2022), which embraces a 'living system' approach. In a similar vein, others have commended to put social justice at the centre of tourism governance (e.g. Higgins-Desbiolles, 2008; Jamal & Higham, 2021; Scheyvens, 2002).

The societal value of cultural heritage, and the importance of preserving it, have clearly emerged from various contributions. Cultural values such as cultural authenticity, cultural sensitivity, education, feminism, solidarity, conviviality and integrity can metaphorically be seen as the 'antibodies' that may prevent the unfolding of a dystopian cultural tourism future. In Chapter 3, as a response to the negative impact of tourism on the natural and cultural heritage of the Sámi people in Lapland, Björn and Lüthje suggest embedding *cultural sensitivity*, as a post-humanist value, in cultural tourists' experiences. Experiences of the Sámi culture, they argue, should be planned together with members of the local Sámi community, and should be based on both human and non-human elements of their cultural and natural heritage. By attending to sociomaterialities, they envision a more sustainable future where culturally sensitive tourism products would promote meaningful interactions between hosts and guests, which, in turn, would foster reciprocal learning and empathy for the planet.

In Chapter 12, *cultural authenticity* is presented as an antidote against cultural commodification. Uysal and Wang underscore creativity as a means to strengthen cultural authenticity in tourism experiences, whereby the risk of cultural commodification would be reduced. The idea that creativity has the potential to revitalise cultural heritage in authentic ways and to stimulate self-development in both tourists and locals lies at the core of creative tourism (Richards & Wilson, 2006). Creative tourism has been recently conceptualised by Matteucci and Smith (2024) as a space of authentic social encounters, co-constitutive heritage experiences, knowledge production and joy. The significance of cultural authenticity, as cultural value, is suggested by other authors in this book. In Chapter 13, for instance, in terms of urban planning and architectural design, Scarciglia questions whether existential authenticity (which relates to subjective inner emotions) should be prioritised over toured-object authenticity. Scarciglia contends that place (toured-object) authenticity will be increasingly threatened by the homogenising force of capitalism. In Chapter 10, Wright associates global connectivity and the widespread use of digital media with the risks of dilution, commodification and standardisation of cultures. The challenge with upholding the authenticity and integrity of World Heritage Sites is considered by Zhang and his colleagues in Chapter 5. Here, the authors note the discrepancies that sometimes exist between different interpretations of the same cultural heritage (e.g. official or governmental narrative versus communities' narrative). It has often been documented that different interpretations of heritage may lead to conflicts between commercial (tourism) developers and heritage conservation communities (Du Cros & McKercher, 2015). For instance, in the South African context, Ashworth (2015: 177) reports that while the apartheid government no longer exists, 'heritage tourism remains largely in a state of apartheid, or separateness'.

The educational value of cultural heritage has been described as a lever that can be activated to resist cultural dilution and commodification. In this book, Zhang *et al.* (Chapter 5) and Šveb-Dragija and Jelinčić (Chapter 8) argue that heritage interpretation can enhance visitors' understanding of local culture at heritage sites and museums, respectively. Furthermore, in connection with tourism, cultural heritage experiences have been linked to the better quality of life of both tourists and locals. In Chapter 12, Uysal and Wang specifically emphasise the role of cultural tourism in supporting local economies, in preserving cultural heritage, and in enhancing community well-being. The positive impact of cultural tourism on rural communities is also underscored by Petrić *et al.* (Chapter

4) who argue that cultural tourism can act as a tool to diversify rural economies, potentially reversing negative trends like depopulation and economic decline. In Chapter 8, Šveb-Dragija and Jelinčić contribute to this discussion by highlighting the role of museums in promoting social welfare and community well-being, and in serving the local community members through transformational activities (e.g. by incorporating health and well-being activities into their programs). The transformative power of cultural and heritage tourism is also underscored by Timothy in Chapter 7.

Bubbles of Ethical Consumption and Governance

In line with our proposition that the future will be marked by a multiplication of spaces of resistance against mainstream cultural tourism practices (Matteucci *et al.*, 2022a) and cultural globalisation (Nijman, 1999; Timothy, 2019), a number of contributing authors envisage that more individuals and collectives will be creating alternative ways of life in the future. For example, in Chapter 6, Farkić insightfully illustrates the development of bubbles of ethical practice in the context of rural communities in central Serbia. In response to fast-paced, commodified cultural experiences, Farkić documents the positive force of slow cultural practices focused on authenticity, deeper cultural engagement and stronger connections between tourists, locals and nature. Her position reflects the view that the 'Slow Movement' is a philosophy of resistance to globalisation (Robinson *et al.*, 2020). She argues that feminist and new materialist thinking will help enact regenerative cultural tourism futures by transcending the boundaries between species, geographies and cultures. Adopting a similar post-humanist approach, in Chapter 3, Björn and Lüthje contemplate the potential for reciprocal learning experiences through immersive cultural experiences in nature. They assert that the entanglement of nature and culture, which is a quintessential feature of the Sámi cultural tourism experience, can help individuals become more empathetic toward the non-human world.

The concept of *community of hospitality* (communauté d'hospitalité), expounded by Wanner *et al.* (Chapter 2), provides another example of the notion of space of resistance. These authors liken the concept of *community of hospitality* to that of *heritage community*. The Council of Europe Framework Convention on the Value of Cultural Heritage to Society known as the Faro Convention (2005) defines the notion of heritage community as 'people who value specific aspects of cultural heritage which they wish, within the framework of public action, to sustain and transmit to future generations' (Article 2b). This definition intimates the idea that individuals and collectives have the rights to cultural heritage in terms of its experience, production and selection. The concept of *community of hospitality*, as *heritage community*, therefore, foregrounds the rights of communities to organise themselves to make decisions that will affect their cultural and socioeconomic futures. *Community of hospitality* is presented as a communalist (or municipalist) approach that promotes direct democracy and local engagement. Wanner and his colleagues provide practical examples, showing how local communities can self-manage cultural tourism activities in a way that benefits them economically, without compromising their cultural values, while still fostering community autonomy, social inclusion and resilience.

A further illustration of space of resistance or heterotopia is insightfully articulated by Wright in Chapter 10. Wright describes a dystopian future in which digital technologies have led to increased surveillance and control over every aspect of private life. Within this dreadful future, the author discusses the emancipatory power of 'counterculture tourists' who seek their leisure and cultural freedom in spaces and places with limited digital comfort, such as forests, woodlands, coastal areas, farmlands and highlands, forgotten villages and towns. Wright suggests that, in a world dominated by digital media and technology, it could be the rebellious behaviours of the marginalised that bring about social change. This view reflects Edgar Morin's (1999: 37) prediction that during the 21st century, various countercurrents will 'intensify and expand into multiple beginnings of transformation'. However, Morin asserts that 'true transformation cannot be accomplished until these currents inter-transform each other, performing a global transformation that will retroact on the transformations of each and every one' (1999: 37).

A number of contributing authors (e.g. Petrić *et al.* in Chapter 4) do not specifically refer to countermovements or alternative collectives; they nevertheless do acknowledge the crucial role that local community members ought to play in planning for sustainable cultural tourism development, and in decision-making. By way of further illustration, in the context of World Heritage Sites in China, Zhang *et al.* (Chapter 5) underscores the value of public participation (residents and tourists) in heritage conservation, such as in interpreting the value of heritage sites. Šveb-Dragija and Jelinčić (Chapter 8) make similar observations in the context of museums. They affirm that, in the future, transformative, eudaimonic experiences will be facilitated through collaborative curatorship involving interdisciplinary teams and a variety of stakeholders from local communities, tech companies and health professionals.

Digital Technologies as Disruptive Forces

The scholarly literature across disciplines has raised important questions around the influence of technologies in advancing human progress. For instance, Nelson and Cooperman (1998) examine what they call the 'paradox of post-industrialization' and note that many commentators tend to be overly optimistic about the prowess achieved in information technologies. While it can be argued that technological innovation allows greater operational control and provides temporary competitive advantage, Nelson and Cooperman remark that not only have technologies introduced a great deal of risk and uncertainty, but also many social problems in contemporary society are due to technological innovations. Therefore, they conclude that if technological advancements continue at an exponential rate, 'social problems likewise should multiply at an accelerated rate' (1998: 593). For instance, a recent review of studies in psychology conducted by Gong *et al.* (2023) on the use of smartphones reveals many disorders associated with the excessive use of such devices. Such psychological disorders include diminished control, distress, low self-esteem, techno-anxiety, depression, digital eye strain, loneliness, decreased affective empathy and sleep disorders among others. Smartphone technology is only one example that many cultural tourists use during their trips.

A significant portion of writings on information technologies in tourism is particularly utopian in praising the extensive benefits of digital tools. In this edited volume, it has become clear that digital technologies will be a disruptive and unpredictable game changer. For instance, in line with Pestek and Sarvan (2021), Mele (Chapter 9) sees great opportunities for tech-savvy Millennials and Generation Z to adopt Virtual Reality (VR) and Augmented Reality (AR) during cultural trips. VR and AR tools are predicted to facilitate personalised, inclusive, 'edutainment' experiences, allowing visitors to interact with virtual artifacts and other users over extended periods of time. VR also presents some opportunities for enabling people to experience places that are impacted by political conflicts, sanitary crises, and natural disasters (Bec *et al.*, 2021). Other opportunities offered by digital technologies are found in other chapters. For example, Šveb-Dragija and Jelinčić (Chapter 8) go as far as to argue that digital environments and technologies will have the capacity to foster transformative museum experiences. This should be no surprise as, in 2006, Sigala had already suggested that the future of museums would be driven by technological innovations aimed at experiential outcomes. In Chapter 11, in addition to the experiential value of digital tools, Gretzel and Sánchez-Amboage explain how digital technologies can make cultural experiences more accessible and available to broader audiences, such as those with disabilities or those who cannot travel. In Chapter 7, Timothy notes a trend towards vicarious travel through social media influencers. He suggests that, in the future, a growing number of 'armchair travellers' will consume heritage through the experiences of YouTubers who will share their immersive experiences with their followers.

Furthermore, a key debate at the intersection of culture and technology revolves around the issue of authenticity. In this respect, various scholars (e.g. Shehade & Stylianou-Lambert, 2020; Mele in Chapter 9 and Petrić *et al.* in Chapter 4) maintain that the future integration of digital technologies, such as VR and digital storytelling, will enhance visitor experiences without compromising the experiential authenticity of cultural heritage. Also, digital technologies will afford multiple heritage interpretations, from authoritative top-down narratives (e.g. AHD) to alternative ones (e.g. marginalised and dissident voices). According to Mele, a further benefit of digital technologies lies in their potential to alleviate popular attractions from the tension exerted by large crowds of visitors (e.g. by creating digital copies). Paradoxically, as Larsen (2006: 247) notes, people increasingly 'travel to actual places to experience virtual places'. While the integration of digital technologies in cultural attractions will probably accelerate in the future, Gretzel and Sánchez-Amboage signal the responsibility of cultural institutions to strike a balance between innovative technological displays and the accurate portrayal of cultural heritage. A significant challenge for local communities, as Richards (2007: 293) indicates, will be 'to create new and authentic forms of culture, which can satisfy the visitor as well as strengthening local identity'. This concern is also expressed by Scarciglia in Chapter 13.

Technologies have many enthusiastic followers. However, some nuanced and skeptical voices are also speaking in this volume challenging techno-utopianism. As Kurzweil (2005) remarks, the march of technological discovery has become uncontrollable and irreversible, causing abysmal changes to human civilisation. Among those changes, millions of jobs are predicted to be impacted by the widespread use of automation and artificial intelligence (AI) across many industries (Tussyadiah, 2020). Another significant threat is what philosopher Nick Bostrom (2019: 465) refers to as the 'high-tech panopticon'. The idea of the panopticon, inspired by Jeremy Bentham, consists of a highly sophisticated digital surveillance system enabled by invasive interconnectivity (or the Internet of Things). The high-tech panopticon bears a striking resemblance to George Orwell's *telescreen*, an electronic device placing citizens under constant surveillance by the state authority. Recalling Orwellian themes, in Chapter 10, Wright envisions a future where society is increasingly controlled by technology, with governments and organisations using advancements in AI and predictive analytics to monitor and limit people's movements. Additional threats would include cultural homogenisation, the trivialisation of heritage (Han *et al.*, 2019), oppressive measures against freedom of opinion and cultural expression, as well as imposed hegemonic values and ideology. As a milder scenario, Gretzel and Sánchez-Amboage (Chapter 11) warn of the new digital divides that might emerge from the rapid digital transformation in cultural tourism. These divides could affect both tourists (in terms of access to and familiarity with technology) and institutions (especially smaller or less-funded ones struggling to keep up with technological advances).

Table 14.1 Key concepts and implications for cultural tourism futures

	Key concepts	Key implications
Governance	**Community of hospitality** as enactments of communalism or municipalism (Wanner et al., Chapter 2) **Sociomaterial teaching** and **cultural sensitivity** as a post-humanist worldview (Björn & Lüthje, Chapter 3) **Community-based tourism** from a new materialist / feminist perspective (Farkić, Chapter 6) **Participatory** (bottom-up) heritage management (Petrić et al., Chapter 4; Zhang et al., Chapter 5)	– Co-construction of communities through the practice of direct democracy, self-management and solidarity – Communal, embodied and place-based approach to cultural tourism – Focus on recognition, respect and reciprocity – Resisting the capitalist value system – Fostering entanglements with traditions, humans, and nature, preserving heritage – Promoting solidarity, care, slow cultural and creative practices – Fostering inclusive understanding of heritage – Leveraging technological advancements for community empowerment and collaborative dialogues
Consumption	**Immersive** and **transformative** cultural tourism experiences (Farkić, Chapter 6; Šveb Dragija & Jelinčić, Chapter 8; Timothy, Chapter 7) **Collaborative** and **creative** experiences (Farkić, Chapter 6; Timothy, Chapter 7; Uysal & Wang, Chapter 12) **Slowness** and **authenticity** (Farkić, Chapter 6; Scarciglia, Chapter 13; Timothy, Chapter 7; Uysal & Wang, Chapter 12) **AR, VR** and **AI-assisted** experiences (Gretzel & Sánchez-Amboage, Chapter 11; Mele, Chapter 9; Šveb Dragija & Jelinčić, Chapter 8; Uysal & Wang, Chapter 12) **Vicarious travel** (Timothy, Chapter 7) **Ordinary heritage** and **pop culture** (Timothy, Chapter 7) **Rebellious behaviours** (Wright, Chapter 10)	– Cultural tourism fosters well-being and contributes to quality of life – Well-being is facilitated by eliciting universal emotions through museum experience design – Well-being and personal transformation are facilitated through immersive, collaborative, slow and creative heritage-based experiences – New types of attractions based on creativity, innovations and well-being will be created – Consolidation of cultural globalisation processes through social media, television and other ICT – Focus on accessibility (incl. in the virtual space) – Digital technologies (AR/VR/XR) will be widely used – Armchair consumers will consumer heritage through the travel experiences of social media influencers – Heritage-based experiences will be enhanced through digital technologies – Escapism from digital society can alleviate humans from controlled states of existence – The travel behaviours of the marginalised may set the masses free

Table 14.1 offers an overview of key concepts and implications for cultural tourism futures which have emerged from the chapters in this volume. The concepts are based on two dimensions: governance and consumption. It should be noted that digital technologies will play a decisive role in both the future governance and consumption of cultural tourism.

Concluding Personal Reflections

Concluding an edited book on the future of cultural tourism is a daunting challenge. It is a challenge because both tourism and cultural heritage are complex, fluid, multifaceted, contextual, and contested social phenomena. The topics and perspectives adopted by the contributing authors only represent a fraction of the critical issues cultural tourism practitioners and communities will be confronted with in the future. In addition, tourism and cultural heritage are both dependent on other larger forces that we dare to simplistically subsume under the general term of the *future of humanity* (Bostrom, 2009). Furthermore, tourism and cultural heritage are both rooted in the past. Tourism history alone is obviously intertwined with human sociopolitical history. The relevance of the past is highlighted by Urry (2016) who argues that exploring the future requires developing an understanding of the ways in which the present, the past and the future are inextricably intertwined. The crucial importance of learning from the past is similarly underscored by Barria-Asenjo *et al.* (2022) who summon that 'we should look at and listen to the past, let history teach us. This will allow us to consider the lessons that, sometimes, seem to be forgotten or are not learned yet' (p. 11).

Through the identification of historical turning points in the evolution of tourism, Yeoman and McMahon-Beattie (2020b) suggest two future scenarios for tourism: 'Degradation – If only we had listened to the past' and 'A balanced future – Learning from the past'. The former scenario intimates a dystopian future, whereas the latter entails a utopian twist. Some commentators have argued that the future is a repetition of the past (e.g. Barria-Asenjo *et al.*, 2022; Hobsbawm, 1995; Yeoman & McMahon-Beattie, 2020a). Combining this argument with the 'If only we had listened to the past' scenario is worrisome. In fact, Westlake (2020), who has examined the past events that led to the construction of Europe, expresses some concerns about likely dystopian European futures. French philosopher and sociologist Edgar Morin (1999) similarly deplores that:

> [w]e have recognized that there are no laws of History leading us ineluctably to a radiant future; we realize that the triumph of democracy was nowhere permanently ensured; we have seen that industrial development can entail cultural ravages and deathly pollution; we have seen that the civilization of well-being can at the same time produce ill-being. If modernity is defined as unconditional faith in progress, technology, science, and economic development, then that modernity is dead. (1999: 36)

If modernity with its certainty of historical progress is dead, as Morin suggests, then by deduction, we can anticipate a new era replete with uncertainty. The COVID-19 crisis, as a recent turbulent event, lends credence to the view that 'the future of cultural tourism may not be so routine' (Richards, 2021: 14). Cultural tourism does not operate in a silo, detached from the influences of the wider political world. In this respect, the rise of conservative politics worldwide does not bode well for artistic and cultural freedom of expression, opinion and speech. It does not bode good days either for freedom of movement. For example, in Spring 2024, scientists, educators, students, and artists (e.g. Ai Weiwei), among others, are increasingly exposed to suspension, censorship and discrimination for voicing their concerns around political matters such as the Israeli-Palestinian conflict. This unfortunate trend is reflected in the latest update of the Academic Freedom Index, which shows that, in 2024, the proportion of researchers worldwide who lack access to academic freedom is comparable to the situation in 1973 (Kinzelbach *et al.*, 2024).

What these recent events demonstrate is that if democratic life is teetering on the brink of crisis, the future of cultural tourism won't be a long quiet river for most of us. Perhaps, as Wright argues in this volume, the fate of cultural tourism will depend on the capacity of social movements to resist the conservative political steamroller that has been fostering the emergence of future 'polycrises'. We concur with Higgins-Desbiolles (2024) who argues that there is urgency to localise (cultural) tourism. Localising tourism (Higgins-Desbiolles, 2024) means letting communities define tourism and heritage themselves. It means empowering citizens as heritage bearers through democratic debates and decision-making (e.g. through deliberating assemblies) about the things that matter to them. Like Wanner *et al.* in this volume, we see much hope in municipalist ideals and initiatives as alternative, autonomous networks of solidarity, which strive to socialise local economies and democratise societies (Thompson, 2021). The examples of Barcelona en Comú in Catalonia, Mondragon in the Basque Country, Cooperation Jackson in Mississippi and Poitiers Collectif in France represent inspiring attempts at developing 'practices and theories of transformative social change' (Thompson, 2021: 322).

As commented by Shvartsman (2023, n.p.), 'what happens next may be terrible or spectacular, or both, but surely it will be fascinating'. Cultures and travel are both beautifully rich and exciting social phenomena. Cultural tourism in its many dimensions represents a positive force that can potentially bring humans closer to each other and to the material world. It is for the sake of a diversely rich, democratic and hopeful cultural tourism future, that individual and collective reflection and debates are desperately needed on such concepts as consumption, culture, heritage, commons, progress, democracy, community and technology. We call, therefore, policymakers, practitioners, academics, students and culture enthusiasts for intellectual openness, experimentation, solidarity and care for the commons.

Looking ahead to a second edition of *The Future of Cultural Tourism*

Editing this volume has been a wonderful journey full of surprises and some challenges. These surprises and challenges are due to our rather loose editorial hand in terms of providing authors with a structure that would have allowed for comparison and dialogue between the contributions. By allowing an open interpretation of what *cultural tourism* is, we gave authors room to be creative, which resulted in an eclectic set of idiosyncratic perspectives on this complex and multifaceted phenomenon. Moreover, our decision not to prescribe a specific time frame of what we understand as 'future' allowed authors to freely explore cultural tourism through their own idea of time, with most of them opting for a near-future perspective. While a short-term perspective may be more easily relatable to our current social life and may provide clearer anchor points to decision-makers concerned with quick solutions to their current problems, a longer-term vision appears more adequate for foreseeing the development of society, identifying social movements, technological advances, and patterns of change into new eras of human evolution (Yeoman & McMahon-Beattie, 2020a). In other words, long-term futures allow speculations and deeper reflections on our social world.

What may a second edition of this book look like? We would welcome more provocative pieces on mid-range and longer-term visions of cultural tourism. Let's say cultural tourism in 2050 and beyond. Longer-term visions would bring about speculative and, perhaps, transformative scenarios, in which both the cultural tourism industry and society would be radically altered. Scenarios might touch upon the rise of new cultural forms, the impact of a global polycrisis, or the emergence of disruptive, new technologies that would redefine cultural (tourism) practices and experiences. A long-term orientation would be crucial for discussing the ethical and social justice implications of cultural tourism practices, thus allowing for a critical examination of how today's decisions would affect cultural heritage, community well-being, social and environmental sustainability in the decades to come. This way, we may stimulate more responsible and forward-thinking approaches to cultural tourism development. Indeed, as Bauman (1976) argues, thinking futures is emancipatory in that it helps us to go beyond the shackles of current routines and conventions. A second edition may expect authors to adopt an orientation that would be either utopian, dystopian or heterotopian. Developing dystopian visions, Urry (2016: 90) maintains, 'act as a warning to those living in the present'. Thinking futures, whether utopian, dystopian or some futures in between, calls many different actors to join the planning table to ensure that future generations will not be deprived from their basic rights to live decently. In fact, thinking (cultural tourism) futures is a call of duty, a courageous attempt to bring sustainability, resilience and ethics on the political agenda of civil society groups, academics, corporations and politicians.

References

Ashworth, G. (2015) Ethnic conflict: Is heritage tourism part of the solution or part of the problem? In Y. Reisinger (ed.) *Transformational Tourism: Host Perspectives* (pp. 167–179). CABI.

Barria-Asenjo, N.A., Žižek, S., Scholten, H., Pavón-Cuellar, D., Salas, G., Cabeza, O.A., Huanca Arohuanca, J.W. and Aguilar Alcalá, S.J. (2022) Returning to the past to rethink socio-political antagonisms: Mapping today's situation in regards to popular insurrections. *CLCWeb: Comparative Literature and Culture* 24 (1), Article 15.

Bauman, Z. (1976) *Socialism: The Active Utopia*. George Allen and Unwin.

Bec, A., Moyle, B., Schaffer, V. and Timms, K. (2021) Virtual reality and mixed reality for second chance tourism. *Tourism Management* 83, 104256.

Bostrom, N. (2009) The future of humanity. In J.K. Berg Olsen, E. Selinger and S. Riis (eds) *New Waves in Philosophy of Technology* (pp. 186–215). Palgrave McMillan.

Bostrom, N. (2019) The vulnerable world hypothesis. *Global Policy* 19 (4), 455–476.

Council of Europe (2005) *Framework Convention on the Value of Cultural Heritage for Society* (Faro Convention), CETS 199. See https://rm.coe.int/1680083746 (accessed March 2024).

Dredge, D. (2022) Regenerative tourism: Transforming mindsets, systems and practices. *Journal of Tourism Futures* 8 (3), 269–281.

Du Cros, H. and McKercher, B. (2015) *Cultural Tourism* (2nd edn). Routledge.

Foucault, M. (1986) Of other spaces. *Diacritics* 16 (1), 22–27.

Gong, Y., Schroeder, A. and Plaisance, P.L. (2023) Digital detox tourism: An Ellulian critique. *Annals of Tourism Research* 103, 103646.

Guia, J. (2021) Conceptualizing justice tourism and the promise of posthumanism. *Journal of Sustainable Tourism* 29 (2–3), 503–520.

Han, D.I.D., Weber, J., Bastiaansen, M., Mitas, O. and Lub, X. (2019) Virtual and augmented reality technologies to enhance the visitor experience in cultural tourism. In T. Jung, M. tom Dieck and M. Claudia (eds) *Augmented Reality and Virtual Reality* (pp. 113–28). Springer.

Higgins-Desbiolles, F. (2008) Justice tourism and alternative globalisation. *Journal of Sustainable Tourism* 16 (3), 345–364.

Higgins-Desbiolles, F. (2024) The end of tourism? Contemplations of collapse. *Journal of Tourism Futures* 10 (3), 476–485. https://doi.org/10.1108/JTF-11-2023-0259

Hobsbawm, E.J. (1995) *The Age of Extremes: The Short Twentieth Century, 1014–1991*. Abacus.

Jamal, T. and Higham, J. (2021) Justice and ethics: Towards a new platform for tourism and sustainability. *Journal of Sustainable Tourism* 29 (2–3), 143–157.

Kinzelbach, K., Lindberg, S.I. and Lott, L. (2024) Academic Freedom Index 2024 Update. FAU Erlangen-Nürnberg and V-Dem Institute. See https://academic-freedom-index.net (accessed May 2024)

Kirshenblatt-Gimblett, B. (1998) *Destination Culture: Tourism, Museums, and Heritage*. University of California Press.

Kurzweil, R. (2005) *The Singularity is Near: When Humans Transcend Biology*. Viking.

Larsen, J. (2006) Geographies of tourism photography: Choreographies and performances. In A. Jansson and J. Falkheimer (eds) *Geographies of Communication: The Spatial Turn in Media Studies* (pp. 243–261). Nordicom.

Matteucci, X., Koens, K., Calvi, L. and Moretti, S. (2022a) Envisioning the future of cultural tourism. *Futures* 142, Article 103013.

Matteucci, X., Nawijn, J. and von Zumbusch, J. (2022b) A new materialist governance paradigm for tourism destinations. *Journal of Sustainable Tourism* 30 (1), 169–184.

Matteucci, X. and Smith, M.K. (2024) *The Creative Tourist: A Eudaimonic Perspective*. Emerald Publishing.

Morin, E. (1999) *Seven Complex Lessons in Education for the Future*. UNESCO Publishing.

Nelson, J.I. and Cooperman, D. (1998) Out of utopia: The paradox of postindustrialization. *The Sociological Quarterly* 39 (4), 583–596.

Nijman, J. (1999) Cultural globalization and the identity of place: The reconstruction of Amsterdam. *Ecumene* 6, 146–164.

Pestek, A. and Sarvan, M. (2021) Virtual reality and modern tourism. *Journal of Tourism Futures* 7 (2), 245–250.

Pritchard, A., Morgan, N. and Ateljevic, I. (2011) Hopeful tourism: A new transformative perspective. *Annals of Tourism Research* 38, 941–963.

Raj, R. and Griffin, K. (2017) Introduction to Conflicts, Religion and Culture in Tourism. In R. Raj and K. Griffin (eds) *Conflicts, Religion and Culture in Tourism* (pp. 1–9). CABI.

Richards, G. (2007) *Cultural Tourism: Global and Local Perspectives*. Haworth Press.

Richards, G. (2021) *Rethinking Cultural Tourism*. Edward Elgar.

Richards, G. (2022) Urban tourism as a special type of cultural tourism. In J. van der Borg (ed.) *A Research Agenda for Urban Tourism* (pp. 31–50). Edward Elgar Publishing.

Richards, G. and Wilson, J. (2006) Developing creativity in tourist experiences: A solution to the serial reproduction of culture? *Tourism Management* 27 (6), 1209–1223.

Robinson, P., Lück, M. and Smith, S. (2020) *Tourism* (2nd edn). CABI.

Scheyvens, R. (2002) *Tourism for Development: Empowering Communities*. Pearson.

Shehade, M. and Stylianou-Lambert, T. (2020) Revisiting authenticity in the age of the digital transformation of cultural tourism. In V. Katsoni and T. Spyriadis (eds) *Cultural and Tourism Innovation in the Digital Era* (pp. 3–16). Springer.

Shvartsman, A. (2023) The brave new generative world. In A. Shvartsman (ed.) The digital aesthete: Human musings on the intersection of art and AI. *Future Science Fiction Digest* 18, November 14.

Sigala, M. (2006) New media and technologies: Trends and management issues for cultural tourism. In D. Leslie and M. Sigala (eds) *International Cultural Tourism: Management, Implications and Cases* (pp. 167–80). Routledge.

Smith, L. (2015) Intangible heritage: A challenge to the authorised heritage discourse? *Revista d'Etnologia de Catalunya* 40, 133–142.

Sterling, C. (2020) Critical heritage and the posthumanities: Problems and prospects. *International Journal of heritage Studies* 26 (11), 1029–1046.

Thompson, M. (2021) What's so new about New Municipalism? *Progress in Human Geography* 45 (2) 317–342.

Timothy, D.J. (2019) Globalisation: The shrinking world of tourism. In D.J. Timothy (ed.) *Handbook of Globalisation and Tourism* (pp. 323–332). Edward Elgar.

Tussyadiah, I. (2020) A review of research into automation in tourism: Launching the Annals of Tourism Research Curated Collection on Artificial Intelligence and Robotics in Tourism. *Annals of Tourism Research* 81, 102883.

Urry, J. (2016) *What is the Future?* Polity Press.

Westlake, M. (2020) Europe's Dystopian futures: Perspectives on emerging European dystopian visions and their implications. *Review of European Studies* 12 (4), 20–31.

Yeoman, I. and McMahon-Beattie, U. (2020a) Introduction: Does the past shape the future? In I. Yeoman and U. McMahon-Beattie (eds) *The Future Past of Tourism: Historical Perspectives and Future Evolutions* (pp. 1–8). Channel View Publications.

Yeoman, I. and McMahon-Beattie, U. (2020b) Does the past shape the future of tourism? A cognitive map(s) perspective. In I. Yeoman and U. McMahon-Beattie (eds) *The Future Past of Tourism: Historical Perspectives and Future Evolutions* (pp. 243–307). Channel View Publications.

Index

Accessibility 13, 24, 172–173, 198, 202
Activism 8
Alienation 7, 231
Architecture 28, 58, 97, 111, 209–210
Art galleries 170–171, 177, 181, 189
Artificial Intelligence (AI) 12, 142, 147, 155, 160–161, 163, 167, 170, 175, 217, 236
Artificial landscapes 210
Artificial villages 52, 54, 63, 65
Augmented Reality (AR) 12, 142, 171, 175, 177–178, 235
Authenticity 53, 60, 62–65, 70, 73, 85, 93–94, 103, 109, 126, 130, 132, 144, 148, 151, 193, 198, 200, 208, 215–216, 232–233, 235
   Cultural authenticity 52, 231–232
   Existential authenticity 13, 109, 144, 193, 207–209, 215, 225–226, 232
   Experiential authenticity 144, 236
   Interactive authenticity 144
   Intra-personal authenticity 207
   Toured-object authenticity 144–145, 207, 225, 232
Authorised Heritage Discourse 7, 11, 73, 145

Bibliometric analysis 190, 201
Bostrom, Nick 1–2, 236, 238
Bottom-up approach 51, 75, 85, 237
Bubbles of ethical consumption 8, 95, 98, 105, 165, 229, 233

Cancel culture 158–159
Capitalism 5, 7, 21, 91, 98, 215, 230, 232
Censorship 166, 239
Climate change 2, 24, 162, 211, 213, 222
Co-creative tourism 109, 112
Collective hospitality 27

Communalism 30, 234, 237
Communities of hospitality 11, 26–30, 233–234, 237
Community participation 58, 230
Community-based tourism 51–52, 237
Conservation 11, 59, 64, 73–74, 84–85, 189, 193, 203, 232, 234
Conviviality 229, 231
Counterculture 13, 155–169, 234
COVID-19 6, 9–10, 51, 59, 66–67, 107, 122–123, 132, 139, 162–163, 171, 174, 177, 180, 191, 198, 203, 238
Creative tourism 5, 66, 112–113, 129, 191, 194, 201, 232
Creative tourist 8, 173–174
Creative traveler 12, 149–150
Creativity 103, 125, 128, 143, 189, 192, 194, 198–199, 200, 202
Croatia 11–12, 54–55, 67
Cultural commodification 232
Cultural dominance 159
Cultural globalisation 114–116, 233, 237
Cultural participation 194, 195
Cultural preservation 193
Cultural sensitivity 11, 35–38, 41, 43, 45–46, 231, 237
Cultural tourism villages 11, 50–67
Cultural tourist 12, 53, 93, 129, 139–152, 173, 179, 182, 195, 200, 220, 231, 235
Cultural values 11, 71, 78, 83, 103, 159, 207, 229, 231, 234
Culture wars 156–158
Cyber traveler 12, 147–151

Dalmatian hinterland 11, 50–67, 230
Debord, Guy 7
Degrowth 3, 109, 115, 216
Democracy 4, 21, 229, 238–239

Democratisation 13
Depopulation 55, 60–61, 64, 97, 216, 233
Digital
  divides 13, 182, 236
  experiences 171
  platforms 145–146, 158
  prisons 13, 155–167
  societies 156, 165
  technologies 12, 139–149, 164, 167, 229, 234–237
  transformation 236
  twins 175, 180
Direct democracy 29–30, 234, 237
Disabilities 182, 202, 235
Disobedience 9
Doughnut economy 216, 218
Dubai 179, 209, 212–222, 226
Dystopia(n) 1–2, 5–10, 12, 52, 71–72, 84, 109–110, 115, 151–152, 155, 161, 165, 209, 211–212, 214, 225, 228, 230–231, 234, 238, 240

Earthly crisis 91–92, 104
Edutainment 143, 235
Emotions 5, 77, 81–83, 122, 124–134, 142, 207, 232, 237
Empathy 232, 235
Empowerment 11, 24, 113, 148, 237
Enhanced traveler 12, 147–149
Escapism 156, 166, 209, 237
Ethical consumption 8, 98, 105, 165, 229, 233
Ethics of care 8, 102
Ethnocentrism 37, 158
Eudaimonia 66
Everyday life 1, 25, 30, 40, 43–45, 108, 124, 159, 174, 208–209
Experiences
  Cultural 5, 53, 93, 107, 148–149, 170–171, 182, 215, 233
  Digital 170–171
  Existential 109
  Immersive 63, 65–66, 110, 112–113, 170–171, 178–179, 233, 235
  Learning 233
  Memorable 45, 110
  Transformative 12, 115–116, 126, 128–133, 234–235, 237
  Virtual 126–127, 132, 148, 151
Experience design 122, 124, 133–134, 237

Faro Convention 27, 29, 227, 233
Feminist/Feminism 12, 92–93, 96, 104, 231, 233, 237
Flamenco 115
Foucault, Michel 8, 95, 229
Freedom 8, 13, 30, 37, 51, 95, 98, 160, 163–166, 226, 234, 236, 239
Future
  of humanity 1, 238
  Dystopian 5, 7, 12, 115, 152, 238, 240
  Heterotopian 8, 11, 98, 115, 165, 240
  Utopian 3, 73, 140–141, 145–147, 150–151, 229, 240

GLAMs 177, 179–182
Globalisation 50, 108, 114–116, 157–158, 226, 233, 237
Global warming 217
Governance 3, 10–11, 19–21, 25, 29–30, 46, 52–54, 58–60, 63, 66–67, 72, 92, 95, 134, 162, 200, 228–231, 233, 237

Health 6, 51, 122–223, 126, 162, 196, 203, 233–234
Hedonic escapism 209
Hegemonic discourse 229
Heritage
  bearers 3–4, 8, 239
  communities 8, 27, 233
  interpretation 84–85, 232
  Natural heritage 11, 70, 72, 77, 82, 102, 232
  Ordinary heritage 109–111, 237
  representations of 144
  revival 141–143
  tourism 12–13, 107–108, 112, 115, 133, 140–142, 189–203, 232–233
Heterotopia 8, 11, 20, 46, 95, 100–101, 113, 165, 229, 234
Hippies 156–157, 165, 167
Homogenisation 6, 230–231, 236
Hopeful tourism 5, 231
Humanity 1, 5, 10, 77, 225

Identity 4, 21–22, 24, 28, 35, 37, 41–42, 64, 94, 97–98, 100, 122, 128, 144, 152, 157, 175–177, 191, 209, 215–217, 225–236

Inclusivity/Inclusiveness 5, 71, 146, 172, 189
Indigenous cultures 34–35, 39
Intangible heritage 145, 147, 228

Joy 81, 232

Lafargue, Paul 9
Las Vegas 178, 209
Localhood 60
Localising tourism 239
Longhushan 11, 70–85

Machine intelligence 1–2
Marseille 19, 22, 28–29, 31
Metaverse 13, 132–134, 170–183
Morin, Edgar 4, 10, 234, 238
Mullis, Kary 9
Municipalism 3, 20, 29–30, 237
Museums 12–13, 52, 58, 60, 63–65, 110–113, 121–134, 143, 146–147, 151, 170–182, 189, 191, 202, 213, 220–122, 227, 232–235

Nationalism 7, 196
Natureculture 92, 96, 103
Natural landscapes 72, 74, 77–78, 81–82, 91, 219
Neoliberal
  logic 11
  order 8, 98, 109, 165, 229
  policies 6, 229
  practices 20, 95
  values 230
New materialism 96, 231
NFT 176, 181
Non-human (elements) 3, 12, 35–36, 42–46, 93, 96, 100–101, 231, 233

Online travelogues 11, 70
Orwell, George 155, 161, 167, 236
Orwellian society 165
Outstanding Universal Value (OUV) 70, 72–73, 76, 82–85
Overtourism 59, 66, 108, 198

Panopticon 236
Participatory 3, 25, 66, 85, 93, 170, 173–174, 178, 181–182, 237

Personal growth 123–124, 126, 132–133, 189, 202
Place-making 4, 194, 201
Pockets of resistance 8, 20, 29, 31, 72, 95, 98, 104, 109, 165–166, 229
Post-digital 159–161, 164, 167
Posthumanism 11, 34–46, 96, 231
Privatisation 6

Quality of Life (QoL) 1, 13, 30, 92, 189–203, 232, 237

Regenerative tourism 5, 51, 67, 139, 231, 233
Resilience 13, 45, 66–67, 116, 234, 240
Resistance 8, 42, 95, 99, 113, 156–157, 165, 233–234
Resort city 13, 209, 214, 226
Responsible tourism 64–65, 92
Robotics 5, 147, 160, 164, 166–167, 217–219, 222
Robots 2, 143, 147, 170
Rovereto 216–226
Rural communities 11, 51–67, 232
Rurality 50

Sámi 11, 34–46, 231, 233
Semantic analysis 76
Serbia 12, 97, 100, 104–105, 233
Slowness 4, 94, 102–103, 237
Slow cultural tourism 12, 91–105, 200
Slow tourism 4, 93, 109–110, 201
Small provincial town 13, 209, 215–216, 222, 226
Smart cultural tourism 170
Social capital 8
Social inclusion 19, 217, 234
Social justice 72, 95, 115, 158, 231, 240
Socio-ecological thinking 92
Socio-materiality 1, 11, 34–46, 228, 237
Spectacle 7
Surveillance 2, 155, 161–162, 234, 236

Tangible heritage 4, 51, 55, 63, 72, 81, 115, 139, 144–145, 147, 150, 189, 195, 228
Taoist culture 75–86
Technoculture 174, 181
Third place 216

Thoreau, Henry David 9
Touristification 6, 110
Transformative tourism 111
Transgression 8, 95

Urban design 207
Urban development 210–211
Urban planning 232
Urban transformation 212
Utopia 1, 3, 9, 45–46, 95, 140, 146, 151, 214

Vicarious tourism 109, 113–115, 235, 237

Virtual Reality (VR) 7, 12, 65–66, 142–151, 161, 175, 177–182, 235–237
Virtual space 127

Well-being
   Community well-being 24, 66, 71, 92, 102, 200–201, 230, 232–233, 240
   Eudaimonic well-being 124
   Psychological well-being 122–123, 196, 203
World Heritage Site (WHS) 11, 70–73, 76, 84–85, 107, 111, 180, 227, 232, 234

For Product Safety Concerns and Information please contact our EU Authorised Representative:

Easy Access System Europe

Mustamäe tee 50

10621 Tallinn

Estonia

gpsr.requests@easproject.com